Always a Brother

By

Sgt Leo H. Brown

90th Infantry Division, 359th, 2nd Battalion, Company F

"Tough 'Ombres"

1 April 1942 – 23 October 1945

ISBN: 978-0692833384

DEDICATION

I express my utmost appreciation to my family for their support and help. My son-in-law Randal graciously interpreted my scribbled account and put it to paper. My daughter Leota edited and helped enlighten the things I wrote by her questions and suggestions.

I concur with President Franklin D Roosevelt, "War is hell!".

I do not recommend it!

CONTENTS

ACKNOWLEDGMENTS

My involvement in World War II (in the European Theater of Operations) as I saw and experienced as an infantry soldier first came about from the encouragement of my nephew Mike Travis. So many World War II veterans are no longer with us and Mike felt that writing my first hand experience would benefit my family and those who are interested in this subject, and to those who are not yet born. It is not my purpose to praise or boast. I do not believe there is any glory in war and I haven't tried to create any. Those who survived can be secure in the knowledge that they ably fought for and supported a just cause. To all the brave men and women who paid and continue to pay the ultimate price for democracy, I humbly dedicate my short account in deep admiration, respect, and appreciation. Perhaps future conflicts will be fought mechanically for the most part. Maybe there won't be much need for the old "blood and guts" soldier of the past. Thank God!
Leo H Brown, 1995

John Schlatter, a military writer, discovered the lost letters that Leo Brown wrote home from his training and life as a World War II infantry soldier on an eBay auction. He did some research, located and informed me about letters written by a World War II infantry soldier to his sister, Zell Chapman in Causey, New Mexico The eBay seller, a serious and conscientious collector, immediately understood and rescued the letters and returned them to Leo's daughter (my wife). A French writer purchased one of the letters written from the hospital in England after he was wounded, and a collector in Luxembourg found the remaining five letters they are all now part of Leo's amazing story.

Randal D. Jentzen
2017

INTRODUCTION

Leo Brown wrote faithfully to his oldest sister, Clara Zell Chapman, during his four years in the army. Two of these years were spent training to become an infantry soldier--the balance of his time was a front row seat to the Normandy invasion and the hell that was to follow.

Zell kept his letters through her final days. She was the first-born and the last to die and outlived her children. The letters were found among boxes that represented Zell's nearly hundred years of life and were sold after she passed, making the journey from rural New Mexico to Anaheim, California.

A military collector and writer from Las Vegas, Nevada, noticed them for sale, researched and located me and asked if I was related to Leo Brown. I told him that I am married to his daughter and knew the man for over twenty-eight years. He informed me about an auction and forwarded the eBay link where I first saw samples of the letters.

The seller and eBay agreed to discontinue the auction and they promptly returned the letters to my wife and refused our effort to pay for them—in all, there were eighty-two, though six letters were already sold and were in Europe. I was successful in getting copies of all six. The French writer has devoted four pages in his upcoming book and the other five belong to a collector in Luxembourg. He said they are mine and joked that all I have to do is come get them.

The initial batch arrived just before Christmas 2016—the first Christmas that Leo's daughter Leota celebrated without her parents—a tough year for an only child who just lost her mother. Leo had passed a decade earlier.

The letters, seven decades old, came to our home and told the story of the man my wife knew as "Dad." He was forty when she was born, eighty-eight when he passed, but now the "twenty-something" soldier, through the letters, introduce himself to her. She was thrilled to see another side to her father.

The young Leo was still the steady, responsible man she had always known, but the younger had perfect vision and hearing, liked watching sports and movies, enjoyed travel, swimming, and he viewed the world with wonder. Something changed him by the time Leota was born and his letters offered insight into his life and complemented the memoirs Leo wrote in 1996, detailing his experiences. I helped with his project.

Leo handed me over two hundred pages that he wrote by hand detailing his life and training in the 90th Infantry Division (the Tough 'Ombres), one of General George Patton's favorite divisions because of their fearless "can do" reputation and willingness to jump into the heat of battle. Leo asked me, "Randal, would you be kind enough to type out my story."

I said, "I would be honored."

His experiences were remarkable. Leo trained harder and longer than many soldiers who had an average of sixteen months of service. Leo was in the army almost four years and was present in the early days of the invasion, was wounded, and three months later stood on the frontline combat. He stayed the course through V.E. (Victory in Europe) Day, occupation, and training for his likely deployment to the Pacific theater of war--the final, deadly assault on the Empire of Japan.

The 90th Infantry Division was reactivated on 25 March 1942 and Leo arrived at Camp Barkeley, Texas, when it was still a makeshift tent city. This would be home off and on for two years leaving for periods of time to train in the California desert, Louisiana, and New Jersey.

The World was at war since the late 1930s. The United States sat on its hands and made a valiant effort to avoid entanglement in world affairs. Franklin Roosevelt, however, advocated for and provided equipment and support to Allied forces, principally, England and Russia who struggled against a common enemy set on their destruction.

America's peacetime military was weak and ill-equipped to enter a two-front war, giving leaders pause. Things quickly changed when Japan attacked Pearl Harbor and was a mistake, as Yamamoto, the Commander in Chief of the Japanese Imperial Navy, purportedly remarked, "I fear all we have done is to awaken a sleeping giant and fill him with a terrible resolve."

America sprang into action and the imperative of every citizen, business (small and large) was to prepare for war. That meant car companies were now making tanks and guns. Industry was retooled to support the effort. Civilians were second priority and it was for duty and pride that young men signed up for military service.

Young women left the farm to labor in the factories and industry, including Leo's sisters Cordia who moved to California and Iva who moved to Arizona. He knew he would see military service but had no idea what to expect: He thought they would train a few days, have rifle practice, and would be sent directly to war—far from true.

I knew the man for twenty-eight years--heard his stories many times, though he never divulged the painful events until he began writing. I suspect he took some of these memories to his grave Margaret, his wife, and Leota, his daughter, recalled that over the years Leo sometimes screamed and thrashed during the night. More than seventy million dead and hundreds of millions of people wounded, abused, and displaced by this protracted conflict left an indelible mark on so many who were swept up in the madness and carnage that was truly a world. War...still there was a glimmer of hope that humanity was not dead but there was a kindness that would eventually prevail. Maybe because man had gone to the brink of madness and there and the will to fight was beat out of the combatants.

Leo was a kind, smart, and a hardworking man. He gave his trust carefully and had no tolerance for small talk. His word was his bond. You could take it to the bank.

My new girlfriend and college classmate invited me home for lunch. I met her mom, but Leo was not home. He was at the cotton gin and would be home shortly. I was sitting in the living room when he drove up in his Dodge pickup outfitted with a propane tank and steel toolbox. When he came through the door, I stood up. He looked me square in the eye, offered me a firm handshake, and said, "Pleased to meet you."

He wore a felt cowboy hat, western shirt, a green down-filled vest, blue jeans, and cowboy boots. He headed to his office and stayed until it was time to eat. He was pleasant enough, but didn't offer much to the conversation. I was twenty, dating his daughter, and imagined him contemplating my death, how he would break my head off—not so. We formed an instant bond.

Leota and I married and over the years spent time at the farm. He enjoyed his grandchildren, Nathan and Rebecca; I could tell from his expression that they pleased him.

One evening, during supper, Leo stood up from the table, pulled his rifle down from the rack above his office door, stepped onto the porch, and watched for a moment, sighted the rifle and started firing! I was astounded!

A pack of coyotes had circled a cow in the pasture, tormenting her. They were about a hundred yards away. He plinked the ground to stir up dust, and successfully sent the pack running for their lives. His aim was remarkable! I didn't know at the time how he acquired this skill.

We "guys" often headed off to his office, and within minutes, we were in World War II. Leo was good with stories, one-on-one.

His Normandy landing during the D-Day invasion was awful and disorganized after their ship hit a mine and sank in two hours. They bogged down in the Normandy hedgerows where the Germans were imbedded and fortified, determined to make the Allied forces pay heavily for their gains. Heated battle raged for the next twenty-eight days, the Allies taking one hedgerow per day and losing a man for every meter gained. On 4 July 1944, Leo was wounded in the assault of Monte Castre, Hill 122. He took a piece of German 88mm fragment that exploded above their heads and lodged in his calf just below the backside of his knee. He was lucky, two soldiers were killed and the others wounded.

A military aircraft transported him to England for treatment and recovery for three months and when he was well enough, shipped back into the war, crossing the Moselle River in flood stages, arriving in a minefield. The next one hundred sixty-one days were relentless, virtually non-stop marching, fighting, often skipping meals, and going weeks without a bath.

Leo was the oldest of four brothers and had two older and two younger sisters. Zell and Leo were the eldest and thus burdened with the responsibility of farming and helping with their siblings at a very early age. Leo went on his first cattle drive when he was twelve, had an incredible work ethic and was driven to finish the job he was given.

The closest community to their farm, Dora, New Mexico, had little more than a hundred inhabitants. The nearest town, Portales, New Mexico, thirty miles away, had just over five thousand. Living conditions in the country during that time were crude with no electricity, running water, or conveniences that are too often taken for granted in the modern world.

He wrote a letter to his sister, Zell, dated 1 December 1944; two weeks before the Battle of the Bulge. "You said, you bet I wish I were there working instead of here carrying a gun. You are just not kidding! The hardest days I have ever seen there would look like a picnic to me now."

Leota and I traveled to Luxembourg last year and walked through some of the sites of the Battle of the Bulge. Beneath the umbrella of trees and inside dense forest one can still see numerous foxholes that remain to this day, and the shrapnel scarred trees, saplings at the time have managed to survive the barrage of German artillery, especially the "tree bursts" that shred the forest and anything in their path.

Leo wrote Zell in late January, at the end of the Battle of the Bulge, that it had been a long time since he sat near a fire and felt warm. One can sense the struggle and the test of will against man and nature as you read Leo's story. One can barely imagine how the serene beauty of the verdant hills of Europe, so gorgeous and inviting were once witness to the bloodiest war in history, nor could one imagine the little old men and women we see struggling with end of life issues were once strong, handsome, and fearless. Many gave up the pleasures of youth in exchange for the opportunity to free the world from Chaos and tyranny. These are too often the unsung heroes to whom we own a debt that cannot be repaid.

STORM CLOUDS OF WAR

"Back in the mid-30s when I was in my teens, I went to Richland, a little school in southern Roosevelt County in Eastern New Mexico. My teacher and a good friend Clarence Fuller taught us history. He referred to events as "the war clouds building over Europe" and said, "It is much more important to know about a war and the threat of war than it is about things that happened in the past."

So he changed the name of our class to "Current Events." He bought us lots of newspapers and anything that he could find about the countries and their ability to wage war and what side he thought they would be on. We found that Great Britain had the largest navy in the world, but most of it was old and predated World War I. The United States was second and most of its ships were also old and most of them built during World War I. Japan had the third largest fleet, most of which was new.

Mr. Fuller thought it would be a long and costly war both in men and dollars. It started a little later than he thought it would. The countries on each side were lined up about like he thought they would be. He couldn't see Russia and Germany on the same side, but neither could he see us lined up with Russia. It would be important which side they were on. I am saying this because I am glad I got involved with Mr. Fuller because I was better prepared for what we got involved in later on. I heard well-meaning people say, "We will whip the "Japs" in 30-60 days."

I was generous-I thought it might take a year.

I wrote this story. I was 23 years old when I got in the Army and 27 when I got out. I am now 78, and I wrote most of this account from memory. If I had written this a few years ago, my memory would have been better. It is faithful to the best of my memory

This is not a document to prove anything. I love this old country of ours, and I hope we would go through this again if we had to. I don't believe there is any glory in war, so I didn't try to create any.

When we registered for the one-year peacetime draft, I was three days too young. But, before the war started, I was old enough, and I was fearful I would be drafted before I could get the crop gathered (we had crops that year). So, I went to talk to the draft board and spoke with the lady in charge.

I asked, "If my number comes up, could I get a deferment until I get the crop harvested?"

She handed me a paper to fill out, but I didn't want one unless it came early enough I couldn't harvest the crop.

So, She said, "The thing to do is wait until you get your call and then ask for the deferment."

That is what I did. My draft number was 1239. Shortly, people were getting calls.

On December 7, 1941, I came through Hobbs, New Mexico about 2:00 P.M. and stopped at a Cafe to eat and people were real worked up. Soon I figured the "Japs" had done something pretty bad. So, I asked what was going on. Little did I know that things would never be the same for me, the people I knew, for America and the world. After that, I would never ask to be deferred for any reason. Little did I know that I was going to Europe.

On December 7, 1941, the Japanese Carrier Task Force made an early-morning sneak attack on the U.S. naval base in Pearl Harbor, Hawaii. The attack was launched without provocation using dive-bombers and torpedo planes. Eighteen U.S. warships were sunk, 174 planes at the air force base were destroyed, and 3581 young men were killed or injured.

Some people with higher draft numbers than mine were getting their calls and wondered how I managed to get deferred. I hadn't. Well, by March 1942 I still hadn't got a call, so I announced to my family that I was going to see what the deal was. My dad said, "You just leave them alone, you go up and start asking and in a week you'll be drafted."

But, I didn't follow his advice. The lady didn't know why my number had been skipped, but she had big stacks of papers on the back of her desk that was to be looked at for deferment. Mine was in the stack.

She said, "You can get 60 days after your call to get your business in order."

I said, "I am just as ready as I will ever be!"

Dad was right. The next mail call, I got my draft notice.

We got a brief doctor's exam in Portales, New Mexico. They charted a bus to Lubbock, Texas. We were put up in a hotel. By bedtime, half of the recruits and the bus driver were drunk. We were called together and told we would have first called at 4:00 A. M. breakfast and start our physical exam at 6:00 A.M. and anyone who was drinking would be shipped straight to Ft. Bliss and the others would have a few days at home. There were about one thousand recruits to be examined.

We stripped and made a pile of our clothes and shoes and went from one doctor to the next. About 12:30 P.M. we were back in our clothes. When we got to the end of the line, one of the doctors showed the other the results of my exam and said, "We need ten million more just like this!"

All checks were positive. We had lunch and waited for the results of our exam and to see if we were accepted. By then, most of them were drunk.

There was a big crap game going. My friend Cleo Ainsworth was broke due to bad luck in the dice game. He came to me and told me that if I gave him five dollars he would give me ten dollars if he lost it when we got to Portales. And if he won, he promised to give me half of the winnings. I told him I would loan him five dollars and he could repay me five dollars regardless if he won or lost. So he took the five dollars and broke up the game that took in the Porter and bus drivers. He was somewhat drunk, and all his pockets were full of money. Later we found out who was accepted and who was rejected. Those of us who were accepted got on the bus to Clovis, New Mexico.

One of the Newberry boys was in charge of us, and he said, "You can eat supper here at Clovis on Uncle Sam."

We were nearly home, so we ate. There was a small cafe in the bus station. We all ordered the big steak and then saw they were big steaks on a big platter. Well, that was a lot of steak for a small cafe. I was sitting with Cleo Ainsworth. With the very able help of the waitress, he got his money straightened out and in his pockets (she may have clipped him a little I don't know). When we cut into our big steaks, the blood ran out. Cleo said, "I can't eat this! Can you?"

I couldn't. He just got up and took one big steak in each hand and carried them back to the kitchen and had them cooked some more. They put them in another platter with fries. We ate, and then we went to Portales. I accepted the five dollars that he borrowed, but no more. Cleo got turned down because he had high blood sugar.

It was a week or ten days we got the notice that there would be a chartered bus in Portales to take us to Fort Bliss (El Paso, Texas). There were several including Jack Altman and Monk Smith (the only ones I can remember now). We got several at Roswell and a few more along the way. In Roswell, we got a big guy with sandy hair. His name was Vernon T. Allen. He was with us all the way.

They didn't try to do anything with us that afternoon. At least half of them were too drunk. The guards just herded them to tents and the next morning. While everyone was sober, they read the riot act to us and told us what would happen if anyone took a drink. Later, we were sworn in the army. Then we got a shot or two and were blood tested, got dog tags, fingerprinted, and they read the articles of war to us. They left us no doubt that we had to walk the line.

I was still in the tent with Jack Altman. He had a lot of dry wit and humor. He said, "Leo, I am all mixed up since they read the articles of war to us."

I said, "What are you mixed up with Jack?"

He said, "I can't remember all the things you get hung for and which ones you get shot for!"

We also got our uniforms that day; I wasn't at all happy with the sizes that I got. Everything was too big.

We heard a lot of rumors as to where we were going. After four or five days we landed in Camp Barkeley, Texas (near Abilene). We got to Camp Barkeley about 9:00 A.M., April 7, 1942 (Leo was in the first wave of 90th Infantry recruits). No one seemed to know where we were to go. We sat out in a company street most of the day. We got moved a few times, but no quarters. Every once in a while, an officer would walk by, and some sergeant would yell, "Attention!"

Well, we would jump up, and the officer would say, "Keep your seat" or "As you were."

I asked a sergeant about it, and he said, "You always come to attention when an officer is present, and it is up to him to say, 'As you were' or 'Keep your seat."

The sergeant also told how to identify an officer by the difference in their uniform rather than wait to see if they had bars on their shoulders. We moved from one company area to another and about dark, it started to pour down rain. We were told to go in a mess hall.

After a while, there were a bunch of sergeants who came in with raincoats on. The First Sergeant who ranted and raved and then cursed us was the most arrogant, hateful, S.O.B. I ever saw. Finally, he got a clipboard with our names on it and started reading them off. He had a German brogue, especially the way he pronounced a lot of names, especially Mexican (many could speak only a little English). He would curse them because he didn't know their name.

When he got 40-50 called, one of the lesser-ranked sergeants took them to another company. After another cursing from the First Sergeant, I thought I would be glad when he sends me with a group and get away from him.

That wasn't to be. I got stuck in his company for the next three and a half years. I saw him a few days before I left France to come home and he hadn't changed: He was bitter at the world and everyone in it. I was the only one assigned to the tent and later a Sergeant Kenen came wearing a very wet raincoat came. We got bedding."

Camp Barkeley, Texas

April 7, 1942

How is everybody by now? I am OK. I landed in Camp Barkeley about 9 A.M. It sure is raining here now. They said it rained all night. It sure is muddy.

I don't know if I will stay or not. I was assigned to this tent but not for keeps. I haven't got any bedding so that I may move on or just move to another tent. I don't know yet.

We get free mail now, so I can't find an excuse not to write. Though, I don't have any address yet.

Things sure are a lot different form Ft. Bliss. One good thing about Ft. Bliss is that it is never muddy. Jack Altman and I were together until we got here and he was sent to the 357th Infantry Reg. I put in for the [cast] artillery, but I didn't have any luck. All of the new guys got K.P. Sat. I guy never knows how clean they run the Army camp until he gets K.P. They scrub the mess hall three times a day. All the dishes are so clean you can see yourself in them. I worked in the dining room. We served 6000 meals that day. I had better close. I have to write on my knee. Excuse the writing. I sent my clothes home if you don't get them, let me know as soon as you get my address. I have a receipt for them.

I did get an address:

P.F.C. Leo H. Brown

Zell, Allen, Darlene, and Don Chapman, Causey, New Mexico Cir. 1943
Aka "Zell and All"

Zell Chapman, Leo's oldest sister, was three years older than he. Leo had three sisters, Zell, Iva, and Cordia. He had three brothers, Delbert, Glen, and Peet. The Chapman's, as well as the Troy Brown family farmed in eastern New Mexico.

Camp Barkeley, Texas
19 April 1945- **Day 12**

Dear Zell,

How is everybody by now? I am O.K. It sure has been muddy since we have been here. I can't leave camp for eight days yet.

The Army is O.K. but I sure wouldn't advise a guy to get in unless he had to. My shoes are 71/2. I have to take two steps before they start moving.

Bye Bye,

Leo

04/26/1942 Camp Barkeley, Texas, Co. F 359th Infantry, US Army (Day 25, Basic Training)

Dear Zell & Allen

How are you all this blue Sunday? I was pretty mad last night. I got in off a gravel haul and all the passes were taken up, the supper was all gone, I went over to the Post Exchange and it was full. Then, I went over to the show. It was full and enough tickets were sold to fill it for the next show. So, I came back to my wigwam and went to bed. I am going over about an hour before the show starts today and try to get me a ticket. The Courtship of Andy Harding is showing. I think it will be good.

I went over and saw Monk Friday night. He thinks he will go home next week and I am going to try but I don't it will do any good. That will be just after payday and every body will be wanting a pass. Fletcher, Judy [sic] in over with Monk and Jack. He said would bring his car back so we would have a way back. I don't think I would be that lucky.

We are going on a 36-hour hike with full field packs starting in the morning after Reveille. I got so that I don't mind the walk so bad—I can always make it all right if the city dudes can.

We had tryout on our mortar gun Friday. I took second place on speed and first on accuracy. There were twenty-four on the tryout. I got it set up and aimed on a stake in 36 seconds—The best was 35 seconds. I had a shot in the left arm that morning. It was sore and stiff. I mean to beat him next time. I beat the number one gunner by half. The platoon Sergeant told me that I would be made Number one gunner next week. I will see about that later...I guess. The number one gunner is next man under the Corporal and if he makes a raise or gets transferred, number one takes his place.

I had a chance to be acting Corporal just after I got here. I had to be a squad leader for three or for months and make it a success before I could get any stripes or raise in pay. I didn't take it because I didn't know any thing about the job then...I wish now that I had taken it.

I think our NCO is going to put in for a transfer to Service Company. I am if I don't do any good here as soon as I get my basic training in.

Well, I will finish the letter. I had to stop for chow. It was pretty good—chicken and ice cream.

I am getting the Portales paper now. Papa had it sent to me. I sure do enjoy it.

Price Gandy and his bunch are leaving today, except the ones that left last Sunday. That will leave the 45th Division empty so they can use a lot more men. I sure would hate to move over there and leave this for the new recruits.

Well, I had better close and go over and try to get a show ticket ahead of time.

Write Soon,

Your Brother,

Leo (or Private Brown)

Camp Barkeley, Texas
12 June 1942- **Day 72**

Dear Zell,

How does this Sunday find you all? I am O.K. It looks like we made some more bad connections after we got back. The folks said you were coming over so I stayed at home. I was afraid if I came to see you, I would miss you on the road over there. I am going to try to come home before Delbert leaves and I will see you then. The telegram got to the orderly room at 2:30 P.M. and at that time I was leaving the main gate of the camp. If I had been ten minutes later I would have received the telegram all right.

It only took us 41/2 hours to get to Clovis. I was pretty blue when I got home and there wasn't a soul there. I finally went to bed and they were there when I woke up the next morning.

We walked 100 miles last week and I gained 6 pounds. Can you imagine that? We walked 18 miles Monday with full field packs. I slept an hour Sunday night. I was pretty blue marching along. They said they had to sweat the 4th beer out of the boys. (Note to reader: June weather in Abilene is 91 degrees with 80% humidity)

I don't know how much sweating it takes, but we have done plenty of it for it was hot. Jack and I went to town last night. We both swore off-- all you can see there is soldiers.

Did you all get your tractor fixed all right? Did you get the crop planted all right? I went up and looked at our crop. It looks good. I don't know what we will do with it though. Delbert gave the cigarette lighter to me in person. It sure was nice.

I had better close and write some more. This makes four and I have four more to go.

Bye Bye,

Leo

Camp Barkeley, Texas
14 June 1942- **Day 74**

Dear Zell,

I am sorry that I didn't write before now, but some how I had the idea that I had already written. I write so many letters that I get mixed up on who I owe a letter and the ones that I don't.

I failed to get a weekend pass after all the kids and Delbert got down here, but I got a pass last night and I think I can get another one tonight. They have been in the camp most of the day. I think Delbert has a better idea what an Army camp is like now.

It sure does look like that Delbert is going to be drafted pretty soon. I sure hate to see him have to go, and it really is going to hurt our business, what little we had.

I got a letter from Cordia real often. She seems to be doing all right. I have heard from Iva several times. I guess they have money in both pockets now.

Somebody stole 7.00 dollars a while back. That sure does hurt the little measly paycheck after we pay our bills.

News is scarce and the kids ought to be back any time.

Answer soon,

Always a brother,

Leo

Camp Barkeley, Texas
27 June 1942- **Day 87**

Dear Zell,

How does this Saturday find you all? I am O.K. we are going to put on a parade for the 90th Division this evening. I sure do dread it.

We had personnel inspection and locker inspection this morning. We were supposed to have the blanket so tight that he could drop a dollar on it and a little ripple would go to each end of the bed. But nobody had half a dollar to test it out with. I bet there isn't ten bucks in this Company. Gambling sure has become dull. I hardly ever see a crap game any more. But, they will be going big about Wednesday, but I don't think we will get our $50 this payday. We may get it about the tenth.

I don't know when I will get to come home but not the 4th of July. I think we will walk over to Abilene and put on a parade for the dear people there. (Note to the reader. Camp Barkeley was 11 miles southwest of Abilene). I don't know why. Most of the people in Abilene wouldn't wipe their feet on a soldier to see so many. They are tired of them.

Has it rained there lately?

That was a nice picture you sent me.

I had better close for it is about time for the parade.

A brother,

Leo

Camp Barkeley, Texas
9 August 1942- **Day 130**

Dear Zell,

How is everybody this Sunday morning? We had all day yesterday and today off. Just my luck, I couldn't get a pass, but I guess it was some body else's time for one. I had a pass the two last weekends. I may get a pass and go to the skating rink this afternoon.

I really do take the shows in. They went down to 15 cents now and by buying a show book, I can see a show for 12 cents.

I am going down to have my pictures taken over, after a while. I have $3.95 invested in pictures and the proofs aren't any good. I guess they are all right but I have four eyes on each picture.

Has it rained there lately and how is your crop? It hasn't rained in so long here. I wouldn't know how to act if it was to rain, and did you say hot?

I guess we are done with our examinations from the inspectors of the 3rd Army. I guess we did pretty well because the Colonel bragged on us and they don't do that for nothing.

We had a big parade Friday afternoon. The Colonel told us that we had finished our 17 weeks of training and we were trained soldiers. I don't know what we will do next.

I guess Roy Capps left all right. I haven't seen him in about a week. Yes, I have been here about a month longer than he has, but he was in the Medics and I am in the Infantry, so we don't necessarily have to go to the same place or at the same time.

Answer soon,

Your brother,

Leo

P.S. Well they called us out on the parade before I got this letter mailed.

There were a lot of people to watch the parade.

Well for the first time in a month we have Sunday off. A couple of boys and I are going to church this morning so I had better dress up a little.

Camp Barkeley, Texas
24 August 1942- **Day 145**

Dear Zell and Allen,

How does this find everyone? I am O.K.

We went out and really got a good soaking this morning. I mean it really did rain. In fact, it rained all day.

There were two Companies that went out to stay three days and started at 4 A.M. today. I feel sorry for the lads for it looks like it will rain tonight.

We are going out Wednesday. I hope the rain slacks up before then.

Jack and I went to town yesterday afternoon and [sic] that is the dullest town that I ever was in! We got back to camp by 8 o'clock.

I have really been seeing some shows here in the last few days. Here are the names of a few of them:

"The Gay Sisters"

"The Lost Horizon"

"Beyond the Blue Horizon"

"Pride of the Yankees"

I saw one last night. I don't know the name of it but it starred Lou Costello and Bud Abbot. It was really funny.

I hope you all have had as much rain there lately as we have here. Everybody I hear from there reports a good crop.

I guess Delbert is just about getting rich now or is he done with his job? Have you all seen the folks since they got back from Oklahoma? I guess that had a big time.

I will have my pictures in the next day or two. I will send one of me.

Answer soon,

Leo

Camp Barkeley, Texas
13 September 1942- **Day 165**

Dear Zell,

How is everybody today? I am O.K. I have forgotten whether I answered your letter or not, so I will do it all over again if I did.

We went out in the hills and stayed all week in our pup tents. It isn't so bad but a raincoat alone is a pretty hard bed. We have a raincoat and a blanket. My tent mate and I cut a lot of cedar limbs for a bunk and we made it all right after that.

They have us take the kitchen out in the field now. They cook up better grub out there. It looks like they could or they would adjust our appetites to it. One doesn't know which.

The second night, we were out there along about 10 o'clock. That night, they called us out and we went out on a compass reading. A boy from Arkansas, a boy from Iowa, and I walked about two and a half miles to a farmhouse and they gave us a slip of paper. It gave four different compass readings and we marched from six to eight hundred yards on each reading. The brush was so thick that we had to crawl most of the time. They left us with the impression that there would be somebody there with a light, but there wasn't anybody there. The idea was to get us lost and see if we could get back. This is where the Arkansas boy came in handy. He was used to the brush. The boy from Iowa kept telling us that we would never get back, but we made it all right and never did use the compass.

They got in the habit of seeing how many of us they could leave off by ourselves and see if we could find our way back.

The weather is a bit cooler than it was before the rains, but it still gets plenty hot sometimes now.

I got the letter from Delbert yesterday. He said we had a good crop and everything seems to be going pretty good.

I got a letter from Mr. Chapman day before yesterday, in fact, I have eight letters to write yet and this is the first one.

No, I wouldn't think Allen would want to work in the broomcorn either. I can't blame him any. I wonder if broomcorn is going to be a pretty good price this year.

Well, did you get the picture I sent you? It isn't so good, but after all, you couldn't expect much.

There isn't any news here, so Bye Bye,

Always a brother

Leo

P.S. I got to be a Private First Class the other day. Tell Allen and Don hello.

Camp Barkeley, Texas

3 October 1942- **Day 185**

Dear Zell,

How does this find everybody? I am O.K.

I guess you know that the folks came down here last weekend, but Papa didn't get to see Mrs. Dooley.

We went to Stanton and saw the Bassets. All the kids are gone. We went back out to the old Baggett School House and the place where we used to live. It looks a lot like it did when we left, but I sure was surprised when we got to Stanton. It isn't any bigger than Elida, if as big.

We have really been taking some long hikes this week. The damned mosquitoes are so bad around here lately that a guy can hardly sleep. They haven't bothered us much until the last few days.

Did you get the barn all built? I guess you have by now.

Monk was due last night. He said he was coming home this weekend.

Zell, I sure do thank you for the stationery because I really use it.

Answer soon,

Always a brother,

Leo

Camp Barkeley, Texas
20 October 1942- **Day 202**

Dear Zell,

How is everyone getting along now? I am O.K.

The boys and I sure enjoyed the cake and popcorn. We all decided that was a good way to pack a cake for it wasn't torn up much.

Iva sent me one of those [Woodbury] shaving outfits. They really did treat it rough. None of the bottles were broken but the box was really torn up.

Cordia sent mea nice billfold. Mine was really worn out but I didn't have much padding for the new one (you know, the folding greenbacks).

I guess it has really been raining there. It rained 7 inches last week. The boys and I went out in it all day and night. I am going to mechanic school. I don't think that I will ever get the transfer because the Company Commander said he wasn't going to train men in the weapons platoon for nothing.

I am really missing some rough hikes and stuff, though.

We all have to make 27 miles tomorrow night. I don't think that will be a snap.

We have lots of work to do on all those trucks but it isn't as hard as Rifle Company.

Have you all got your barn built yet?

Tell Allen and Don hello.

Answer soon,

Always a Brother,

Leo

Camp Barkeley, Texas
6 November 1942- **Day 219**
Dear Zell,

How does this find everyone? I am O.K. I feel pretty good—just got in out of the woods this afternoon and got a shave and a shower, clean dry clothes, and a haircut. Now, I will have a dry bed to sleep in tonight.

It really has been raining out there this week. I guess it rained every day.

I am still working as a mechanic. We really do have a lot of night work to do and the trucks for they use them all day. But, I still like it a lot better than following a rifle company.

I went up and tried to get a three-day pass tonight, but the Company Commander wasn't there so I didn't get one. I don't know, I might not have gotten one anyway. I would have done some pitiful pleading with him.

News is scarce, so answer soon.

A brother,

Leo

Clipping from Newspaper attached to letter (1942)

"This Army Pay Just a Waste of Time"

A Chinese trainee in MRTC apparently figured it was just a waste of time to go through the procedure of getting paid. Soldiers were signing the payroll the other day when the top sergeant discovered the Chinese private wasn't on hand. The missing lad was located at the bathhouse doing his laundry. The sergeant asked the soldier to accompany him to sign the payroll.

"Me no sign," was the reply. "Me wait hour to sign last month. Me wait hour to get pay on payday. Then me lose money in crap game in 30 minutes. Me no sign."

Camp Barkeley, Texas
6 November 1942- **Day 219**

Dear Allen,

Well how is everything going with you? They are not so hot with me at the present. I went up and asked for a three-day pass and the Company Commander wasn't there so the 1st Sergeant put me on table waiter for tomorrow so I for sure couldn't get a pass.

Has it come a killing freeze there yet? I guess Delbert is done cutting our grain by now if it hasn't been raining too much.

It really has been raining a lot here this week.

I am still working with the Service Company as a mechanic—sure as hell do wish I were transferred over there. I like that outfit a lot better than I do this.

Did very many go deer hunting around this year?

They moved a bunch of new rookies in the Company next to us this week. I sure do feel sorry for them when they go around with their arm swollen up after the shots and for standing in line with raincoats and shoes while it is raining about daylight for [peter] inspection. This goes on for about two weeks.

Answer any time you can,

Private Brown

90th INFANTRY DIVISION

The 90th Infantry Division was activated during World War I. Arriving late in France, it played an important and vital role in the drive at Saint-Mihials and in the Meuse-Argonne Offensive.

It was re-activated at Camp Barkeley, Texas on March 25, 1942 under the command of Major General Henry Terrell, Jr.

In 1942, the 90th Infantry Division wore an olive drab patch with the red inscription, which in past days stood for "Texas" and "Oklahoma." This was because it was originally made up of almost exclusively men from these two states. Later, the division drew its men from every state in the United States, and the "T-O" came to represent "Tough 'Ombres" (Tough men).

The 90th was reorganized on the triangle pattern; the division consisted of the 357th, 358th, and the 359th Infantry Regiments. Next were the 315th Engineers and the 315th Medical Battalion as well as the 415th Quartermaster Battalion, the 90th Signal Company, and the 90th Reconnaissance Troop Division Artillery commanded by Brigadier General John E. Lewis. It consisted of three 105mm Howitzer Battalions, the 343rd, 344th, 915th, and the 345th (155 mm Howitzers).

On January 23rd, Brigadier General Jay W. McKelvie assumed command of the Division, and in March, the unit moved to its staging area at Camp Kilmer, New Jersey. There followed a brief period of final physical preparations before the March 23, 1944, departure from New York Harbor and, later, Europe. But first, the reformed 90th had to undergo the long and arduous job of becoming a well-trained unit. That is when I arrived.

During the next thirteen weeks, or what they call "Basic Training," we learned to do close order drills (marching), the manual of arms, and gun inspections. And we had a night or two in the field. They had to "learn" those city boys what it was like out in the dark with the same kind of food and the same kind of exercise. The overweight ones lost weight, and the underweight recruits gained.

I mentioned earlier that we got Vernon T. Allen on the bus at Roswell. He wound up in my tent. I liked Red (that was his nickname), but he was a slow thinker and too old and still 20 pounds overweight. When we went to a class or any place for instructions, he always depended on me to tell him about it and what we were expected to do. He had said that he had cooked in a cafe all his life and no way would he do that here. Well, by this time we were doing some obstacle courses, and Red came in all skinned up and so sore he could hardly get around. I found out there was an opening for a first cook that would be same as a buck sergeant. They could have a day on and a day off and a lot safer job than what we were doing. He could see he wasn't going to make it there, so he applied for the cook job and done it until the war's end. When we went through the chow line day or night, Red always took good care of me. If it were dark, I would speak to Red, and he would dig a little deeper. I could always get a loaf of bread or some ground coffee if I needed some.

When I first joined Company F, they told us that our platoon sergeant was in the hospital and would be for a good while. A sergeant that was supposed to be in charge of the machine guns was acting platoon sergeant joined us. He was old regular army, about 45, and looked like he was about used up. There was quite a lot of static between the draftees and the old regulars. A lot of them were alcoholics, and in peacetime, it was a sure grocery bill and a little loose money. But, the word went around that our platoon sergeant was homosexual, and he disappeared. No one ever mentioned that he was leaving or why. Then, we got Weldon Jones, a real nice draftee.

When I got to Camp Barkeley, the 90th Infantry Division had 17,000 men. The commander was a Major General (2 stars). A Brigadier General commanded the artillery (1 star).

I put in my entire time in the 359th Infantry F. Company, which was a rifle company. It had three rifle platoons with a weapons platoon, two 30 caliber machine guns, and three 60mm mortars. I was with the mortar section, but most of our combat time, we played the part as a rifle platoon. Much of the time, we didn't have enough men to carry that much equipment. We usually kept our mortar ready to fire flare shells. A flare pops out a real bright flare attached to a parachute to let it down real slow. They were chiefly used to light up German tanks so ours could get a shot at them at night.

1942 was a dry, hot summer. We had a lot of rain in the spring, and little or none until September. When the division had been built up to full strength, it ultimately consisted of men from Mid-America: from Texas to the Canadian border, then other central states like Minnesota, Montana, Louisiana, and New Mexico. Little states, like New Mexico and Wyoming, didn't have many men, because there weren't many of the original men left; there was one from Albuquerque named Frank Savergella and myself.

There was one Jewish boy from New York. We had a few Jews; one man's name was Goldwasser. When they came in they were issued tin mess kits; we got the old C.C.C. (Civilian Conservation Corps) aluminum mess kits. One day, Goldwasser asked, "Brown, why is it my mess kit rusts and yours doesn't?"

I told him, "I came to camp, I washed and dried it good, and rubbed it good with gun oil."

Well, he tried it, and it didn't rust, but on the other hand, he didn't need a laxative for a few days! I told him, "I forgot to tell you; you should wash the oil off before you use it!"

You had to be on your toes, or someone might play a trick on you!

There was a big ranch west of Camp Barkeley known to us as Hankins Ranch. We had a lot of maneuvering. There were lots of cedar trees for shade, but they had a million chiggers. The soil was shallow with lots of rocks and digging those foxholes was tough. There was a lake south of the ranch. It was the only water near. We did lots of beach landing and simulated river crossings on it.

I was so surprised that we were fooling around so long when we needed to get over there and whip those Japs. I had no idea we would do KP and drill and do parades. I thought we would clean up a rifle and practice a day or two and go "Jap hunting." We followed the old army training routine like we had for a hundred years. Of course, it takes a while to equip and ship that many soldiers, even if they are already trained. I just couldn't get over that many grown men out marching along the roadway when there was so much to do. I never did consider that I had never trained an army.

When we got up in camp, no matter how black dark it was, there was always this fellow yelling and selling papers. One morning during the summer of '42, he was yelling, "Soldier's pay has been raised to 50 dollars per month!"

Boy, did that sell papers! There were big black headlines about the Battle of Midway Island and the sinking of four "Jap" aircraft carriers. But by breakfast time, no one had found where we were going to get 50 dollars per month. Finally, someone found it! There were about four lines over in the paper where some senator had proposed that the soldiers should be getting 50 dollars per month, but it did sell papers! Later on, we did get 50 dollars per month!

In taking different tests, such as setting up a 60mm mortar and zeroing it in, firing a rifle and reloading it, or making a pack, I was never more accurate than anyone else, but I was always the fastest one. One time, we had a full field inspection on the parade ground. It was a battalion size (1,000 men). When the inspection was over, the Lt. Colonel said, "When I say 'strike tents' (which meant to take down the pup tents and make field packs), the first man standing up with a full field pack will get a three-day pass starting—now!"

Well, I didn't suppose I would win, but I was going to give it a try. I didn't look to see how everyone was getting along. I just worked on mine and stood up with a pack on and looked around. No one else was through! A sergeant came over and inspected my pack as said it was O.K. The colonel said, "He (me) will get the three-day pass."

It was the only three-day pass.

I got the pass and got a bus ticket. The bus was about full, but I found a seat. There was a soldier just behind me, and the only seat next to the back was room for two people. A black woman was sitting in it. The soldier told her to get up and sit in the back seat. The back seat passengers said it was full (it looked full to me). The bus driver came back and told the soldier, "If you want to ride, you have to sit with that woman."

The soldier came unglued all of a sudden. He grabbed the driver and said, "I bet you are hired to drive a bus and not to fight, and if you don't want to fight, you had better drive!"

All of a sudden, there was plenty of room on the back seat for the black woman.

We changed buses at Sweetwater, and I noticed the soldier got on the Lubbock bus as I did. We got to Lubbock and had a long layover. There was a travel bureau there. This soldier asked me where I was going. I told him that I was going to Clovis, New Mexico. He said that he was going to Friona, Texas. He wanted to get to Farewell, Texas. The towns were near each other. He asked me if I would be interested in going to a travel bureau to rent a car if we could go pretty quickly. I told him I would if the price wasn't too high. So we went in, and the man in charge said they were getting the car greased up and they would be ready to go. So we gave him our four dollars each and waited a few minutes and no car. So he called and said they were checking the tires. The work went on until it was nearly bus time. I said, "We might just as well get our money back and get on that bus."

The guy said, "You can get on the bus, but you are not getting your money back!"

The soldier picked up a pop bottle and said, "You take this and see that shoe shine boy don't get on my back. I am going to mop up the floor with this guy."

He reached across the desk and got him by the lapel of his coat and jerked him about two-thirds of the way over the desk. The guy quickly changed his tune. He said, "Oh please let me go. I will be glad to give your money back!"

And he did!

The shoe shine boy threw up his hands and said, "I'm not going to give you any trouble."

So we got our money and didn't get in jail as I thought we might. When we got on the bus, the soldier told me that he generally didn't act this way, but they were doing these things because they were dealing with a soldier and knew they couldn't afford to get in trouble. After the return from visiting home, I was back at camp.

Our mean first sergeant was still the same. He was always yelling and swearing at us. There was nothing we could do about it. After breakfast, the first sergeant would call us out to police the area. We would line up shoulder to shoulder with plenty of yelling and swearing. We would pick up all the cigarette butts, bottles, etc. Then, the officers would take over for the day's training.

One morning, our platoon leader, Lieutenant Lepene, got there a few minutes early and stood around the corner where the first sergeant couldn't see him. When the first sergeant got through cursing at us out, Lieutenant Lepene stepped around the corner, and he cut the first sergeant down to size. He said, "If I ever hear you swear at these men again, I promise you that you will be a buck private before night."

That was music to our ears!

When we first moved to Camp Barkeley, we lived in permanent tents. They had plywood floors, a four-foot wall, and a tent on top of that. It was a lot better than no plywood. There was a gas heater in the middle and a light bulb on the top but no windows. We could roll up the tent flaps when the weather was favorable. One time, we went out in the field for a week. When we got back, all the tents were gone, and they had moved three of the floors together and built what they called huts. They had two windows in the back and one on each end and two doors in front. They were mainly plywood and felt paper, but they were a lot better than what we had.

I made a lot of good friends. One was Odell Sulivan. All in all, I guess he was the best friend I ever had. Odell was from Oklahoma. There was also Ralph Duncan from Arkansas.

Some time after we got raised to $50 per month, someone decided we all should have $10,000 in life insurance. I was not much of a believer in insurance. I had a $3,000 policy. Nearly everywhere we went we walked. And that wasn't very dangerous. Everyone took out the $10,000 except a few, including myself. The rules were, we had to meet with our platoon leader and let him (second lieutenant) explain the need for having the extra insurance. The next week it was the company executive officer, then the captain, then the major, then the lieutenant colonel, and finally the full colonel. By the time I got the lecture and got back to my company, all of the passes were gone, and the word was that we or I was the only one who didn't get the extra insurance. It was always on Saturday afternoon. When this full colonel got done, he wanted to know if I had any questions (he was really nice). I asked if I had a choice or if I would be continually harassed and wouldn't get a pass until I took the insurance. He said, "You're damn right you have a choice; that is the reason we are fighting this war. Buying this policy doesn't have anything to do with whether or not you get a pass."

I told him," It's been six weeks since I've had a pass."

He said, "You'll get one today. I will call your first sergeant, and he will have your pass ready when you get there."

He kept his word, but he didn't like it.

When we finally got some free time, Camp Barkeley was pretty boring. To keep things interesting, someone was always pulling pranks on someone else. It was Sunday evening, and Odell Sullivan and I were in the little town of Merkle (about 15 miles west of Abilene). We were going to the bus stop to catch the bus to Abilene when we ran into our buddy Ralph. He was drunk as a dog. We tried to talk to him, but he could only mumble. We knew we had to get him back to camp or he would be in trouble. Ralph was a good soldier and was never absent without leave (AWOL). Sullivan and I were about the same size, and Ralph outweighed us by about thirty pounds. So, we each got on opposite sides of him, got one of his arms around our neck, grasped his belt in the other hand, and managed to get him to the bus stop and on the bus. It was the same deal at Abilene to get him on a camp bus. Then, we had several blocks from the bus stop to our hut, but we finally made it. We sat him on his cot, undressed him, hung up his clothes, laid him in bed, and pulled the blanket up under his chin. He raised up and laughed like crazy and said in a perfectly normal voice, "I just wanted to see if I ever got so drunk I couldn't get back to camp, that you two would help me."

It was all a put on! He was no more drunk than we were. But, one had to do something to keep from going nuts!

We got a new company commander about this time, Captain

Quincy L. Brown. Well, one day we came in at noon, and there were three or four letters on my bed. I opened them and started reading them. I got to one that didn't make much sense, so I liked to see who signed it and didn't know them either so I had another look at the envelope and it was to Captain Quincy L. Brown. I figured he would kill me at least. I rehearsed the best speech I could think of and went to the orderly room and stood at attention and started on my speech. The good captain said, "Don't you worry about it a minute. If things had been reversed, I would have done the same thing."

I decided that we had a good captain.

In the late fall of 1942, I got my first furlough – ten days plus traveling time. Well, it expired on Saturday night at midnight. We never had reveille on Sunday morning or any formation. So, I left for Clovis on a bus about midnight, which was customary.

When I returned, I went to the orderly room and turned in the furlough, but when I got to our company, I knew something was wrong. I didn't see anyone. The Charge of Quarters said they had first call at 4:00 A.M. and everyone went camping where an armored division was stationed. The C.Q. told me there would be a truck leaving at 2300 that would get there the next morning. Odell Sullivan came into our hut and said, "Hey Brown, where is everyone?" then he said, "I am AWOL."

I told him, "Me too!"

We got down there (Brownwood, Texas) at about 9:00 A.M., and that night, the captain called us up and told us we had the right to call for a general court marshal or take the punishment that he would deal out. He said a court marshal would go on our record and the punishment would not. So, we took the punishment, a week of extra duty. We were out in the field all week and went back to camp Barkeley on Saturday. We had to do a little work Saturday evening and wait tables on Sunday. So, we got off real light.

I didn't have a serious girlfriend, so I didn't worry about getting married on furlough. A lot of them did. It was just the thing to do, and I am sure most of them got married for the right reasons. But, a lot of them didn't work out. They found several on the East and West Coast, where a lot of soldiers and sailors were shipping out. They found a lot of girls who were married to 15 or more soldiers and sailors. They got big bucks out of the allotment checks. That was before computers and everyone had to use their social security number.

We did a lot of firing on the rifle range. I did manage to make sharpshooter but had a lot of guys made expert, and a few made a possible score (in other words if they fired 20 times, they got 20 bull's eyes).

One was Eddie Love. He always got a great score. But when we fired rapid fire, I was always first to get through. We also shot the 45 revolvers (this was only for mortar and machine gunners). We fired six times, and I got five bull's eyes and one near the bull's eye. The platoon sergeant was telling another sergeant that Brown wasn't facing the target correctly and his arm wasn't straight enough, but when they read the score, and I made 59 out of a possible 60, he said, "He is from New

Mexico. He probably carried a gun most of his life."

I don't remember the score, but we did well with the 60mm mortar, and I could still set it up ready to fire quicker than anyone else could.

We were taking some twenty to twenty-five-mile hikes by this time, and it was sweltering. We only had one canteen of water per day (one quart), which wasn't near enough. One day we were on one of the hikes, and we had stopped up on the side of this hill with no shade and no water. It was about 3:00 P.M. Down at the bottom of this hill there was a windmill and a huge steel tank full of water. There were concrete troughs for cattle to drink. White-faced cattle were-were standing and laying around. Well, that made us thirsty. There were the officers just a little way from the water tank, but they weren't between the water and us.

I said, "I am going to go get me a drink. Does anyone want to go with me?"

One of my companions said, "You can't get water. They will run you off."

I said, "I am thirsty and don't have any water. If I get some, I am better off, and if I don't, I haven't lost anything."

When I started walking toward the well, Odell Sullivan said, "I'll go with you!"

On our way to the well, I told him, "When we get there, we'll drink all that we can hold and then fill our canteens and give it to the guys too chicken to go. But if they make us pour it out, we will still have a belly full of water."

Well, we drank all we could and filled our canteens, and no one bothered us.

When we left they noticed us, and here came the second lieutenant.

He asked us, "What do you mean getting that water?"

I told him, "We were thirsty."

He said, "Don't you know that it might be poisoned?"

I said, "If it were poisoned, those cattle would be dead."

Well, he hadn't thought of that, so he said, "It's unauthorized water so pour it out!"

We did, but we still had a belly full of water, and the second lieutenant had shown us his authority too.

Somewhere about here, someone with higher authority decided we should be a motorized infantry division. I had a history of some mechanic work. I went to a

government-sponsored school about a year before this. There was another guy, Clyde Sperry. He was about 35 but had worked in a shop for ten years or so. He and I were sent to work at Service Company but still stayed in F Company. After two months, they found that they didn't have enough trucks and de-motorized us, and I became an infantry soldier again.

Clyde Sperry stayed. He was too old for infantry and had lots of experience as a truck mechanic. That was a much better and safer job than a line company. He got along well for about eight or ten months and got drunk and cussed his second lieutenant out. So, he got busted. I was still on their books, so they sent word for me to move to Service Company. No one asked me if I wanted to or not. It was an order to me from our first sergeant. I should have taken it, but I didn't like his attitude or anything about him. So, I said, "I went over there once and was sent back."

He said, "Oh, you have got it that they don't have anyone else."

I said, "You have a damn good man already, and I suggest that you keep him, and I will stay where I am."

All in all, that was a dumb decision. Service was a safe place, but toward the end of the war, a lone shell came back and hit Service Company, and Clyde Sperry was killed. I had a letter from his wife thanking me for sending his personal things home. I don't know who did because that was the first I knew about it.

In the summer, while the weather was real hot, everyone had to make a twenty-five-mile hike in a maximum of eight hours. They set a day for our battalion. Well, we got lucky, and the clouds were low, so it was cool. We had this one guy who had a sprained ankle, but Captain Lepton told him he would have to start, and it was no use going very far because if you didn't make the full twenty-five miles, you had to do it all over. So, this guy went a mile and a half and fell out, and in a little while, the Battalion company came along. He explained his problem. The Battalion company told him to unroll his pack and put his blanket around him because he had been walking at four miles-per-hour and would be sweaty and might take a cold. But, the soldier kept telling him how it would be OK, and that he didn't need a blanket, so the commander ordered him to unroll his pack and get his blanket. He did, and instead of a GI blanket, he had a round, Quaker Oat box in his pack. He got some extra duty.

We had fifteen minutes for lunch break, a sandwich and a cup of water. I then quickly changed my socks; it was important to keep your feet dry, so we always carried an extra pair of socks. We had a relatively new replacement from the New England states, but I can't remember which one; he was real small and was forty-two years old (too old for the infantry). We had made sixteen miles, and John Cuna (that was his name) said, "I can't go any further!"

I told him, "You're just going to have to make it because if you fall out, you will have to do it all over!"

He said, "I'm going to fall out!"

So, I took his pack, and Ralph Duncan took his rifle (we had a five-mile break every hour). We kept threatening him, and he made it. My shoulder, which had the extra pack on, felt like it was going to fall off. I didn't sit down the last two or three five-mile breaks. But, John Cuna made it, and we made the twenty-five miles in 6 hours and 15 minutes!

We had several parades during the hot weather, and a lot of those men had never been in that kind of heat. When we had to stand at attention for very long, a few of them would turn a white, pasty color and fall over. We had an old colonel reviewing the troops, and he said, "It is just their imagination!"

Of course, it was a heat stroke. Anyway, he was about old enough and trying to make another grade before he retired. Whether it was real or not, he fell over one day and was dead at the time that he hit the ground! Of course, there was another parade for his funeral.

I caught guard duty. I was at a road crossing near the chapel where the funeral would be held. The Sergeant of the Guard told me, "At this given time, there will not be anything but military troops on this road that goes in front of the chapel. Then one hour later, there will be no traffic at all."

After the non-military traffic was to be stopped, here came this important looking guy in a Packard car.

He was in charge of the Base Exchange at the camp. I stopped him. He was very unhappy! Finally, he said, "Do you know who I am?"

I said, "I don't care if you are President Roosevelt, you are not going down this road."

He said, "I don't believe you can stop me."

I said, "Bet I can put eight bullet holes in this Packard car!"

So he turned around and stopped and said, "You will hear about this!"

I never did hear about it again.

At the next hour when I was to stop everyone, here comes the division commander, a Major General (two stars) in his Packard car with a flag with two stars waving. So, I stopped them and gave him my best rifle salute and told him that I was sorry, but there was not to be any traffic until after the funeral, and he would have to take another route, which I pointed out to him. I would take them back to the chapel, and he said, "Excellent, soldier!"

He did as I told him. Well, when I finally got off of guard duty, I told the sergeant that while on guard, I got to stop the Division Company General Terrell.

The sergeant said, "Oh my God! You don't stop a general!"

I told him, "You said at the specified time that I was to turn everyone back." He said, "I thought you would have enough sense to let the general through! Boy, you will hear about this!"

I said, "You might, but I won't because I carried out my orders."

We never heard a word about it. But if I had said, "No one is supposed to go through at this hour, but being a general, I guess it would be all right." I would have had trouble.

With the 90th being made up such a large area, there was many customs and sayings with some, and the one I am talking about was from Iowa. Just in common talk, he would refer to the person he was talking to as you S.O.B. I had told him more than once," That is fighting talk where I come from! I would rather you not refer to me as an S.O.B."

I was laying on my bunk with no shirt or shoes and in this guy walks and refers to me as an S.O.B. I called his hand, and he grabbed me by the feet and jerked me off the bed and said, "What are you going to do about it you S.O.B.?"

I grabbed him by the feet and down we went on the floor between two bunks and I came out on top. I banged his head against the floor until he had enough. After that, we got along okay.

1942 was drawing to a close. We were coming to an end of a training schedule, and lots of rumors were going around. Some that we were moving others the 50% would get furloughs for Christmas, and the other half would go overseas. The strongest rumor was that we would go to Louisiana for maneuvers. For some reason, that I don't know, we got the word on the bulletin board that there would be no leaving for Christmas. No one went home for Christmas except for my good friend Odell Sullivan, and he went AWOL. He was so good at coming up with a story and good reason that he only got his wrists slapped and a week of extra duty. After this, he enjoyed cooking up a big lie that you just had to believe. The next summer, he got a three-day pass and took four extra days. Odell had been AWOL at least four times.

We were out in the field on maneuvers when he got called up. There sat the lieutenant colonel on a log, and he had several court marshal books stacked beside him. This put the fear in Odell. The colonel said, "Private Sullivan, I am going to tell you something, then you can tell me your story, and knowing you, I am sure you have a good one. But, if you go AWOL anymore, I will promise you that you'll know what the inside of a prison looks like." Then, he continued, "I want to hear your excuse for being four days late."

Odell said (I sneaked up to the kitchen tent so I could hear), "Sir, I simply do not have an excuse. My mom, little sister, and little brother are still out on the farm. They have some crops, but the fence is all bad, and they can't keep the cows out. So, I decided I would stay and help them fix the fence to keep the cows out of the crops. That is all that they have to live on. I knew it would get me in trouble, but I can take it a lot better than my mom and the rest of the kids can." Then he stopped and held up his hands and said, "I don't have an excuse."

The colonel stood up and said, "Damned if I don't believe you again! But, it still goes that the next time you will go to prison!"

He made a believer out of Odell.

In 1944, when we were on the east coast, and we all got furloughs before we were to go overseas, Odell got a furlough the next day after I did, but he lived closer, so we were to get back the same day. He slept in the bunk next to mine. When I returned, I got back three hours early at 6:00 P.M.; I had until 9:00 P.M., but there was no Odell. I woke up at 12:00 A.M. and his bunk was empty. At first call at 5:00 A.M – no Odell. So, I went to talk to the platoon sergeant and asked if he knew anything about Odell and he said, "Oh yes, the Red Cross called and got him a three-day extension because his brother was coming home on leave from the Navy the day Odell had to return. So, the company commander granted him three extra days."

The platoon sergeant, Weldon Jones, was a good man.

I said, "But he doesn't have a brother in the navy!"

Weldon said, "I know that and you know that but the company commander doesn't!"

Well, by the time 1942 was over, we realized we had a long and dangerous war to deal with. The Marines had won some minor battles in the Pacific, and the navy won a major victory when they sank four Japanese aircraft carriers in the battle of Midway. After Pearl Harbor and Germany had declared war on the United States, it gave Britain and other allies hope, and they were ready to put an all-out effort on the Western Front that had come to destroy the German ground army.

We at Camp Barkeley were having inspections and parades and watching training films. We were in between things. A lot of us thought we were getting ready to ship out, and we still thought it would be the Japs that we would fight, for they did bomb Pearl Harbor. But, I and everyone else knew that Germany was just as much of an enemy.

Camp Barkeley, Texas
29 November 1942- **Day 242**

Dear Zell and All,

How does this find you all? I am O.K.

I got your long letter yesterday and enjoyed it very much. We are near Yuma, Arizona now, close to the Mexican border.

Tomorrow, we will take off on foot to Needles, California. It is almost the entire length of the state of Arizona. I think we average 20 miles per day and in the sand we will slide back about 10. So, that makes 30 miles. Ha Ha

Zell, I wrote Cordia three letters since she wrote me. She was going to get married and she never has answered at all. She must have got mad at something I wrote, though I tried not to make her mad. Just between you and me, I think what she got mad at me because I knew the guy she was going to marry and I think she intended for me to convince you all that she was doing the right thing, but I just couldn't see it that way at all and I said so. But, I didn't intend to make her mad.

I met this guy but I don't know him. I also met his brother who is in the Navy and I also met the Navy boy's wife. They are all from Virginia and are all southerners.

Today is Thanksgiving for us. We had turkey, candied nuts, and fruitcake. It sure was a good dinner.

I guess we will be in the desert for Christmas from the way things look now. But our maneuvers are over the 11th of December.

Time is Short and I have a lot of letters to write, so write soon.

A Brother Always,

Leo

Camp Barkeley, Texas
26 December 1942- **Day 269**
Dear Zell,

How does this find everybody and how was Christmas with everyone?

I got the present you sent me today. They feel O.K. and they are something I can use when they wear out I can salvage them and get some new ones for free.

I sent your present to the folks. You may have already received it. I don't know.

I didn't have any boxes or anything to mail them in.

I sure do hate to see the folks have to move down the sand hills but I guess that is the best they can do this coming year.

I guess Delbert is practically in the Army by now. I sure do hate to see him go, but I guess he will be relieved for he as been about to go for six months.

I can't understand that L.D. Wells. I thought he would really learn a lot when he gets in this Army.

How is Allen and [Wart] getting along with their work?

Write soon,

A brother,

Leo

LOUISIANA

Before January 1, 1943 was over, we knew we were going to Louisiana by truck convoy. It took us two days. We camped near a little town in East Texas. We got a pass, but there wasn't anything to do or see. We—Odell Sullivan, another guy, and I—had the idea we would get us a good meal. But, that was not to be. Every place was full of soldiers. We went to a bakery and couldn't get in reach of it. But, I handed this lady a dollar bill and asked her if she
could hand me something to eat if she had anything to eat. I got lucky and got a cherry pie. So, we ate it there. We heard later that next day you couldn't get a little bit to eat because it was all gone, not even a loaf of bread.

The Air Force got word of my tests some way before the maneuvers of 1943. My eyes were better than 20/20, my hearing was good, my aptitude was better than most, and my size would just fit in any aircraft, (either you can qualify to fly combat aircraft, or you come back to the infantry), so I went and took all the exams and did well. When I had finished all the exams, the examiner told me to come down the next day and bring my birth certificate.

I said," I don't have one."

He said, "When you get one, you will ship out to Shepherd Field."

I finally received my birth certificate and got it just before we shipped. There was no way to get a transfer then. The next day we landed in Louisiana. The main thing I can remember about Louisiana is it rained most of the time, and it was always muddy. We were wet and muddy, and it is surprising how much heavier a full field pack is if it is wet. Then, for about half of the two-month maneuvers, we had a horseshoe roll of our overcoat, and it was always wet. Finally, we got them dried out and put them in a truck.

I remember the fleas, chiggers, and ticks. We straightened a big safety pin, used a cigarette lighter to heat it red hot, then stuck it to the ticks; they turned loose so not to leave their heads buried in our skin. We had to take turns removing ticks from each other.

We were along the Sabine River. It was pretty cold, but we took several swims that would get rid of the fleas and most of the chiggers—especially if we used a lot of G.I. soap. Part of the time we were in Texas near Jasper, and the town in Louisiana was Lidwili. At one time, it had been rich in yellow pine lumber, but it had all been cut out and sold; no one had replanted any pine trees, so they were destitute people. The land was poor, sandy soil; I guess if they had put a lot of fertilizer on it, they could have grown something. Mostly, there were two-room shacks with moss hanging from the roof to the ground, and if they had a car or truck, nearly all of them were Model T Fords. However, they had plenty of coffee and twist tobacco. Late in the afternoon, these guys would come out through the woods with a three-gallon bucket in each hand and sell us coffee for 25 cents for a canteen full. I don't know what they sold the twist tobacco for; it was too rugged for me. They grew the tobacco and cured it out and made it into a twist.

I understood that maneuvers were to train the officers on how to move troops and supplies. What we were doing made no sense to us. There were a lot of high-ranking officers there. There was a three-star general in charge of the maneuvers. One night, he came to a bridge across the Sabine River.

It was only a couple of miles to his headquarters. There was a soldier guarding the bridge, and he stopped the general's command car and asked for a written pass to get across the river.

The general didn't have one, so the private said, "I am sorry sir, but I will have to turn you back."

The general said, "Now listen, private; it is only two miles to my headquarters, and I will have to travel about 50 miles to cross another bridge."

However, the private stood firm. The general said, "I am in charge of the maneuvers, and I sign all the orders!"

Still, the private wouldn't let him across. He said, "If you sign the orders, you above all should know you had to have a pass to get across."

So, the general and his driver took the long way around after getting the soldier's name and serial number. The next day, he hunted the soldier's commanding officer and told him, "That is a damn good man you have guarding that bridge. I want him made staff sergeant before the day is out."

We had a month of maneuvers and a three-day rest and got a shower and washed our clothes, and there was a lot of crap games, and a few soldiers somehow got some booze and got loaded. It was a miserable time, but Camp Barkeley looked a lot better when we got back to good showers and a mess hall with tables and benches.

We got another general over the training, and he said, "The only way to be sure a bunch of men can walk 35 miles per days is to walk 35 miles per day."

We did this for quite a while. Then we got a general who had been over in the Philippines. He took charge over a significant area of our training.

He said, "You have damn good men, but you are going to kill all of them!"

So we started walking 9 miles in one hour and 45 minutes in the morning while it was cool. We watched training films the rest of the day. We had a lot of training and films on the fortified lines with all of the concrete pillboxes and "dragon's teeth" (this was known as the Siegfried Line). Well, those were in Germany, and we were sure we were going to Japan. That was why we went to the swamps of Louisiana—we thought.

I drew guard duty at the stockade. They checked out two prisoners to each guard. There was one fellow from our company in the stockade. His name was Pierce. He started out to be a good soldier, but he got an unexpected pass home that his wife didn't expect either. She was out with a second lieutenant. Well, he took the rest of the week to try to straighten out his marriage. So the company commander was good to him and only gave him a week's work after hours, which wasn't much. There was a big mean sergeant in charge, and he had a real long clipboard (about 3.5 feet). Pierce came by when his name was called. He barely moved his head to me, and that sergeant tried to hit him in the head with that clipboard but missed him. He cursed him and said, "You S.O.B.! I ought to kill you!"

So, I guess their rules were pretty terrible there.

The way Pierce got in the stockade for six months was he refused to finish his week's extra duty for being AWOL. He quit work, and the first sergeant went to put him to work. He referred to the first sergeant as a "Nazi S.O.B." He wouldn't work for him (I am sure that the first sergeant could have handled it better and got it worked out, but he didn't). Well, the company commander went and told him, "You have had a bad experience, and you just lack two hours of having your week's extra duty put in, so if you finish that, we will forget your run-in with the first sergeant."

But, he refused to follow the company commander's order and got a court marshal and six months in the guardhouse. You will hear more about Pierce later on. He got in great trouble.

I got a couple of guys who had a sob story like Odell Sullivan could cook up. One, his mother had died; he couldn't get a pass. The other one said his wife had a baby and he couldn't get a pass. I didn't believe either of them, but I didn't think they were criminals. When we signed for them, this mean sergeant told us to either bring them back or take their place until they are caught. I told them I wouldn't be mean to them or overwork them, but one way or another I would bring both of them back, and they both said they understood that I would.

They were shoveling coal. Part of the camp (not our part) was using coal. I let them rest ten minutes out of each hour. I found out later that prisoners were not supposed to get a rest break. Before the day was over, they wanted me to let them go to the Base Exchange. Maybe you have to be tough, but I just had them one day.

By this time, all the rumors had run their course. We found out that we were going to the California desert. Well, the battle of North Africa was over, but I think we were building up and training a lot of infantry divisions and they didn't have anywhere to put us. They asked for volunteers to go and help get things set up. We all volunteered, but Stanley Skay was the only one in our platoon to go. He said we didn't know what we missed. They worked every day in 135-degree heat to set up the tents; there were no floors, lights, latrine, and no running water. At one end of the camp was the company latrine, and a cook tent was at the other end. We had several alerts and dry runs on how to board a train. When we had an alert, it was treated as an emergency; if anyone left the camp, the punishment was severe. Well, Odell Sullivan had a date in town, and he announced to me that he was going. I tried to talk him out of it, but that was a real challenge to him.

Somewhere in the Hills of Louisiana

16 February 1943- Day 321

Dear Zell,

How does this find everyone? I am O.K. I am kinda tired of writing letters for this makes ten have written today.

We are done with the first phase (14 days) of our maneuvers. Now we are supposed to have five days to clean up, get paid, and clean our equipment. Between a pine knot fire, a GI can, and a creek, I washed my clothes this morning. They don't look the best, but they will do, I guess.

We kinda have to forget some of the finer things of life. A bale of hay to make a bed out of would look like an innerspring mattress to me.

On the 14-day "problem" all the nights were a blackout. We eat supper after dark and breakfast before the daylight. When we eat and have to sandwiches, an apple or orange for lunch. Once they enemy captured our chow truck and two days we were cut off from the rest of the regiment and chow truck. We didn't get anything to eat except two sandwiches they managed to bring up to us.

I think the bread must be some that they brought from Camp Barkeley; It was so hard, but you didn't hear a growl out of anyone. We didn't have any blankets those nights. I bet I pulled enough grass to make nine bales of hay. I didn't sleep so poorly. I had quite a bit of hay in my mouth, eyes, and hair the next morning but I got it cleaned out.

We captured five Companies of the enemy and made them do without chow with us. That didn't make us feel so bad, and we captured nearly all their artillery and trucks.

When the 14 days were over, you couldn't count the prisoners, but now they have them back, and we will have to whip them again.

I am going to try to go to town tonight and get my "ears lowered" as "Wart" says and eat the biggest T-bone steak thy have (and I mean a big one).

I had better close for you probably are getting tired of this bull.

Answer soon,

Always a Brother,

Leo

P.S. Tell Wart to write.

Louisiana
Co. F 359th Inf., A.P.O.-90
29 March 1943- **Day 362**

Dear Zell,

How does this find everybody? I am O.K. and really happy indeed because the maneuvers are over and we leave tomorrow for Barkeley. I will really be glad to get there where we can sleep on a bed and take a shower once in a while.

I haven't seen a bed of any kind except a bunch of grass, pine needles, or leaves since we left there and they are always wet.

I wish you had part or all of this rain we have had here for the last week. All the creeks got up and we stayed wet for about three days and nights without taking my shoes off and I haven't slept without my clothes since I left Barkeley.

I am sending the pictures back but I want you to send me the negatives as soon as you get this letter. Send them to:

P.F.C Leo Brown
Co F 359th Inf. A.P.O.-90
Camp Barkeley, Texas

I don't know anything to write and it is getting dark and I am leaning back against a pine tree so there.

Answer soon,

Always a Brother,

Leo

CAMP BARKELEY
PREPARATION

Camp Barkeley
Co. F 359th Inf., A.P.O.-90
17 April 1943- **Day 381**

Dear Zell,

How does this find everybody? I am O.K.

Zell, I will pay you for the pictures when I come in on that furlough that I hope I get pretty soon.

Boy oh boy I could go for some of that fried chicken you were talking about. I haven't had any complaints on the chow since we got back to Barkeley, though it isn't home cooked food.

I hope you all made out O.K. without Glen. I am sure you will make out better than Papa would have without him. In fact, I couldn't see how he would make it at all. He told me that he liked to work for Allen just fine.

I am glad he went back home for if they call him for the Army I believe he would have a lot better chance to stay out than he would have otherwise. Don't you?

If it doesn't rain pretty soon neither of you need any help to farm. I didn't guess. Will you?

Have you seen the show "Hit Parade of 1943?" It was really good. Also, "Hello Frisco Hello."

Delbert said he had been to Nashville. He may join the Grand Opera. Ha Ha

Write soon.

Always a Brother,

Leo

Camp Barkeley, Texas
Co. F 359th Inf., A.P.O.-90
27 April 1943- **Day 391**

Dear Zell and All,

How does this find everyone? I am O.K.

Was it very hot Sunday? It was 109 degrees here (43C). It hasn't been very hot since then.

I went to see Monk. He was on K.P. Boy, was he sweating. He said he thought he would get furlough the 4th of next month. I don't know for sure if I will or not. I am getting anxious to get a crack at some of the fried chicken.

Jack Altman is in the hospital. He hurt his back on maneuvers and has been hospitalized for two weeks and said they gave him one aspirin, that was all.

Delbert said I had better get a furlough the first chance because he didn't know when he would get one. If I knew for sure, I would come.

I would try to wait and get one the same time, but you can't tell about this Army. I was supposed to come home then 18th of this month and we went to guard the railroad for President Roosevelt when he went to El Paso and I didn't get it. I was in Austin for five days.

I never did expect to see this place again when I left.

It's lights out, so answer soon.

Love always,

Leo

Camp Barkeley, Texas
Co. F 359th Inf., A.P.O.-90
2 June 1943- **Day 427**

Dear Zell and All,

Hello everybody. How is every little thing in New Mexico? I guess everything is O.K. here.

Zell, I really could use that Chicken supper with the ice cream and pie now.

I got a letter from Delbert. He is in Maryland now. He said he just barely made it back in time. I got back here Thursday at 4:30 P.M.

I got my stuff straightened out and laid down on my bunk and decided I would rest a little while before Chow. I woke up the next morning for first call with my clothes and my shoes on and very hungry.

The Company was all out in the field. I thought I would have to go out the next day, but I didn't, so I slept all day. So, I kind of caught up on my sleep.

Saturday I caught K.P.

Answer Soon,

A brother,

Leo

Camp Barkeley, Texas
29 July 1943- **Day 462**

Dear Zell and all,

How is everybody in New Mexico? I am O.K. about the same as usual. It has really been hot here but is cooler here today because it is cloudy and looks like rain. It hasn't rained here in a long time.

There was a girl I used to go with came through Abilene yesterday afternoon. She had been to Texas visiting her folks. She called me and I went down and went to Sweetwater with her. She had a 5-hour layover there. I got back to camp at about 4 A.M. I was pretty sleepy at first call at 5 A.M.

It sure seems good to see someone from your hometown. She was really anxious to get back to New Mexico. She said it was too hot here.

I got Cordia's picture. It was really good, but I guess I will have to send it home. I try to stay ready to leave all the time but it looks like it is all in vain.

They are checking all our clothes and equipment. There is talk of moving out, but that is nothing to get excited about because we have done all of it before.

I had a letter from Delbert and he was about ready to go back there.

No, the pictures were not so hot, but I still want some of them as you say, it was hard to tell at first what was wrong with them.

A brother always,

Leo

P.S. Tell everyone hello for me.

Camp Barkeley, Texas
Co. F 359th Inf., A.P.O.-90
13 August 1943- **Day 499**

Dear Zell and All,

I received your letter and testament. This is really O.K.

I guess I have read it more so far than I ever did all the rest together and I promise it will go wherever I go.

I had two letters from Ray yesterday. He says they have been doing a lot of fighting for the last six weeks but they were whipping the "slant eyes" every day.

I am going to try to get a three-day pass and come home before we leave here, which isn't very long off.

It is still hot and dry here. I hope you have had a rain there before now, but I doubt if you have. Is your crop still holding up pretty well?

I had a letter from Papa yesterday—that makes two or three in the last 161/2 months. Not so bad. Eh?

Iva said Dale was in 3A. Now that doesn't look so good. I sure hope the folks can keep Glen out of the Army and I believe they can.

News is scarce. So, write soon.

Always a Brother,

Leo

P.S. I forgot to send your pictures. I am going to have a devil of a time getting the other one from Ray Altman. He has it and is in New Guinea. I forgot that you wanted them back. I will send the tow I have but I couldn't promise about the other one.

Camp Barkeley, Texas
Co. F 359th Inf., A.P.O.-90
26 August 1943- **Day 512**

Dear Zell and All,

Well, how does this find everyone? I am O.K.

It is still very hot and dry here and guess it is there also. I hope you get some rain.

I got back to Camp an hour and a half before first call. I was sleepy at first call. I didn't feel so good for the 41/2 mile run but I made it O.K.

I am sending your picture back. I would have brought them with me but I never thought of it.

Is Allen about through plowing now?

Have you and Don got those grapes canned yet?

I had a letter from Delbert yesterday. He seems to be O.K.

News is scarce, so I will drop you a line when I get to my new home.

Bye for now.

Leo

P.S. I have been prison guard at the stockade yesterday and today. It is easy but I don't like the job.

Camp Barkeley, Texas
Co. F 359th Inf., A.P.O.-90
1 September 1943- **Day 518**

Dear Zell and All,

I sent my watch to you yesterday. You can give it to the folks the next time you see them.

Has it rained there yet? It hasn't here but it has been a little cooler the last few days.

I saw a really good G.I. show this morning "Secret Agents."

Answer Soon,

Leo

He said, "We won't leave tonight!"

He was unaware that a family member had died and the Red Cross was trying to contact him.

I was the first person they contacted. I told them he had a friend that was in the 357th Regiment that he might have gone to see, but I didn't know what company. I told them it was just as likely he went to see a show. They went to the theater in camp and stopped the show and couldn't find him. They called the 357th and couldn't locate his friend. About 11:30 P.M., he came in. I alerted him that he was in trouble, and whatever he told them had better be good. He went and called home.

He told me, "It isn't illegal to be out of pocket as long as they can't prove I left camp."

He told them that he had seen a show and that his friend was being transferred to another unit, so he just walked around the camp until bedtime. He stayed with his story, and no one believed him, but they couldn't prove otherwise.

The next day, I was in the supply room which joined the orderly room, and I heard the first sergeant ask the company commander, "Are we going to write Sullivan a three-day pass to go to the funeral?"

The captain said, "No, I will go to the regiment and get him a seven-day emergency furlough. If we give him three days, he will stay the extra four days anyway."

I came back and told him. He thought that was funny, but he realized he had about used up his luck and got back on time.

Before we went to the Arizona and California desert, the 359th Regiment (about 3500 men) was sent to guard all the railroads near Austin, Texas, to the Mexican border. I was with a group that was guarding the bridge on the south edge of Austin. They brought us food, but it was real cold by the time we got it. There was a cafe about a quarter of a mile from there. We took turns eating and standing guard until our money ran out. Then the G.I. got food, so it tasted better. We had no idea why we were guarding the bridge. One night, there was a railroad detective who joined us. He didn't know why he was there either, but the next morning, we got a newspaper and found that President Roosevelt had been to Mexico City to discuss the war with the Mexicans.

CALIFORNIA
DESERT TRAINING

We finally boarded trains and went to the California desert. It was 135 degrees under a roll-up tent the day we got there. We didn't get out in the sun except to eat. The next day, we marched one hour. We kept building up each day until we were able to march all day and night. It had become cooler by then. The advanced group that volunteered to go work had been there about six weeks. Old Stanley Skay was burned to a crisp and had lost about fifteen pounds. He was in the machine gun section and was one of the best machine gunners I have ever seen. He was with me V.E. Day (Victory in Europe). He was a big guy, like 225 pounds with round shoulders and long arms. He always looked real sloppy. He was to be depended on, and I did.

All the prisoners were turned out of the stockade that had six months or less. That included Pierce. After we had gone to the desert, it looked like he was going to straighten up and make a soldier again. I think he would have, but the first sergeant rode him all the time. He couldn't forget that Pierce called him a "Nazi S.O.B."

We landed in the desert south of Needles, California, on August

29, 1943. We would go on a three-week maneuver on one side of the Colorado River and to the other side the next time. We moved against the 93rd Infantry. They were guarding this pass in the mountains. We were 39 miles south of them, and that was our starting place.

We could take off to engage them anytime. So, we had early chow (about 12:00 noon) and took off, and we fought them the next morning at about 4:00 A.M. We had walked all night and had covered 39 miles of the worst desert sand I ever saw. The umpires in charge of the exercise gave us a clear victory because they were all sleeping except very few. They didn't take life too seriously. They wanted us to carry their weapons because they were now prisoners of war. We went near Yuma, Arizona and stayed for a while (the reason we didn't leave Yuma was that the trucks were stuck in the sand). They had to get winch trucks to get them out (it looked like a desert storm).

We were given a water bag to every fifth man, so water was a better deal than the summer before. Some soldiers were lost with only one canteen full of water. They died, and some heads got bumped together among the big shots. After that, things improved.

We got a few passes and one three-day pass to Los Angeles. I went to see my sister, Cordia, each time. I borrowed eighty-five dollars from her and her roommate while I was out there on a three-day pass (that was back in the days when men paid the bills). I never was much of a poker player, but I thought, "If I lose, I will still owe the eighty-five dollars! But, if I win, I am back even again."

We were out in the sand for some firing problems for two

weeks. I played a little poker in some mid-sized games. I drew my first straight flush and won a few dollars. I won the eighty-five dollars, so the first chance I got, I sent a money order to Cordia so I wouldn't lose it.

The food was poor while we were on desert maneuvers. A lot of the time, it was straight C rations. There was one little can of meat and vegetable hash that I could never eat. It had such a terrible smell. I couldn't take it. There was one called "Meat and Vegetable Stew." It was a little better but not much. There was one just about like pork and beans, three round contained round crackers, and some powdered coffee, and three lumps of sugar. They were not very good. It was hard to figure any way to carry them.

We went on a five-day problem of compass reading. The mortar section, which I was in, was attached to a rifle platoon. This officer was in charge. He was a lawyer from Dallas, Texas. He was one of the nicest and smartest officers I ever met. Anyway, they would take rations and water out in the desert and bury them. Then, they would give us a map and compass and some physical landmarks such as a hill or a road. Then there would be a compass reading to a stake with a white piece of cloth about the size of a handkerchief tied to it. From the starting place to food and water would be about four miles with a change of compass reading about four to six times. We would start as soon as it got dark and stay with it until we found the food and water.

This first lieutenant never missed a one; usually, in an hour and a half, we had our work done. We would sleep the rest of the night. I would heat some water and make some coffee and have crackers. At noon, I had the pork and beans. I was hungry enough that I ate the meat and vegetable stew, but never the meat and vegetable hash. It wasn't much fun lying out in a pup tent in that kind of heat. It was much better than being out in the sun or hunting food and rations as some were still doing. Never did the lieutenant miss one. There were a lot of guys who came in sick. I guess they must have eaten the meat and vegetable hash.

When we had the kitchen truck for breakfast, we had powdered eggs. If Vernon Allen (Red) was on duty that day, they weren't too bad. Back when we had bacon, Red had saved the bacon fat, and he would season the powdered eggs. He could fix up something that a person could eat. The other first cook didn't care. He did the minimum. Those old powdered eggs floated around in yellow-greenish water that cooked out of them were pretty terrible. When we had pancakes, Red mixed his up the night before, and they would be real good and tender. He would stand and cook them until everyone had enough or until he ran out of batter. The other cook would make three per man, and the rest went into the garbage pit. Three of his pancakes was about all a man wanted.

There were big mountains to the south and a little bit to the west. Well, I had been around mountains in the western United States enough to know they were a way off. A bunch of those guys said, "We're going to get the cook to make us lunch, and we are going to walk to the mountains where it is cool."

I told them that it was probably seventy to eighty miles, but they got their lunch and a canteen of water and walked until noon. The mountains looked the same. So, they ate their lunch and turned back and walked back to the camp.

We covered a lot of miles. We typically went for two or three weeks, sometimes into Arizona and sometimes in California, but the terrain in both was just the same.

Soldiers unfit for infantry duty were weeded out. They were sent back east. I don't know what they did with them. There were always men and boys that should have been left at home. When we got replacements from another unit, they were usually some that they wanted to get rid of. Once, we went to Yuma, Arizona and walked to Needles California. I don't know how far that is, but it seemed like a long way with a lot of sand!

I had always read and heard that there was water in the desert if you knew how to look for it. We walked it for three months, and I never saw any sign of water. There was underground water. It was pumped in irrigation pipelines to the desert units like ours. We had showers that seemed like a good idea in the daytime, but that night it was cold. We finally got a coal-fired water heater and, about November 15th, we got some little heaters in our tents but didn't get fuel. We could go out in the desert and get sagebrush and stuff to burn. Then, we would rip off a little coal that was used for the water heater.

One day, we went out early in the morning in trucks. We had light packs, one ration, and a canteen of water. No one, including company officers, knew where we were going. We went quite a long way, and we started walking back to camp. We had a five-minute break each hour and fifteen minutes at noon. There were no more breaks except these. As the day was breaking, we got to camp. Later in the day, we found out that it was an endurance march, and we all made it! No, we didn't have the 42-year-old (John Cuna) anymore. When we left Camp Berkeley, he was tearing tickets at the theater at camp. With all of the kids that he had, it cost the government more than the company commander. We never heard about him.

We went to Arizona for another week and came back on Christmas Eve. Desert training was over. We had a good Christmas dinner. Four or five guys had a bottle of booze and got stoned. Other than that, it was a very peaceful Christmas. We had a few inspections and heard a hundred rumors as to where we were going. But, I think we all knew that the training was over.

The 30th of December, it rained all day and all night. We had to fold our canvas cots and turn them in, leave the tent spotless and the little stoves knocked down and clean. We had quite a bit of fuel, so we were keeping the tent hot. We didn't loosen the tent stakes, and it ripped from the side all the way to the top. Just before we left, we would have had to pay for it, but it was midnight when we left, and no one knew about it. We had to bury what fuel we had left and set the little stove out in the rain to wash it often.

California Desert
Co. F 359th Inf., A.P.O.-90
8 September 1943- **Day 525**
Dear Zell and All,

I arrived in the California desert this A.M.

This place is about like the desert between Alamogordo and Hot Springs, New Mexico.

We have a tent city here and camping cots—the folding kind. It could be worse, I guess,

Write soon,

A brother always,

Leo

California Desert
Co. F 359th Inf., A.P.O.-90
15 September 1943- **Day 532**

Dear Zell and All,

How does this find everyone? I am O.K. I guess considering every thing.

We have been here a week and no one has gone over the hill yet. That is a pretty good record for the desert.

We haven't had it very tough yet. We are trying to get used to the heat. We put in 8 hours a day. We start with four hours of hiking and a little of everything in the afternoon.

We have had quite a bit on desert navigation. It is interesting but a very complicated subject. It is about the same was as they navigate a plane or a ship at sea but grid lines, longitude, latitude on the map with a compass. That is the only way to keep from getting lost at night, especially on the daytime for that matter for one pile or sand and gravel is the same as the other.

I saw a lieutenant today that was out looking for a bunch of the 358th.

He said there was one whole Company and a bunch who were lost and another Battalion (900-1000 men) also lost. I guess they have found them before now.

There was a platoon that got lost before we arrived here and 7 or 8 of them died and they located the others with a patrol plane.

There were 7 or 8 tanks lost out of the 8th Armored Division. Last night there were many planes out looking for them.

I don't know if they ever found them or not. So, you can see a guy man need a little desert navigation before he starts on maneuvers.

It gets from 125-135 degrees here every day. It was 135 degrees the second day we were here and that is in the shade of a tent.

There was a refrigerator truck out of L.A. that brought about 15,000 pints of ice cream out here yesterday. At first he wouldn't sell but one to each person. There are only 3500 in this Regiment. So, you can see he had 4 to 5 per man. Then it began to melt, so he sold us all we wanted. I ate 4 pints. It sure was good! It was the first ice-cold thing I had tasted since we have been here.

We come through Clovis on the way out here and stayed an hour and took exercise all the time. They wouldn't let us go to town at all.

Has it rained there yet? I bet it really is dry if it hasn't. I haven't gotten a letter from home dated later than the 1st .

Answer soon.

A brother,

Leo

California Desert
Co. F 359th Inf., A.P.O.-90

20 September 1943- **Day 537**

Dear Zell and All,

How does this find everyone? I am O.K.

I guess you have heard from me a couple of times before now. I wrote you a card the first day I was here. They were just slow to send out the mail.

I think I will get a three-day pass to L.A. in about a month or less. They are going to run a bus from this station to L.A. and back for $6.00. That isn't so bad.

I called Cordia's phone yesterday. She had just gotten out of bed. It was about 10 A.M.

It is only 112 degrees. It seems plumb cool.

By the way, I got a good conduct medal Saturday. What do you know about that? It was for over a year of good conduct in the service.

We can go to San Bernardino for a weekend pass but I never have gone yet. I went to the Colorado River (51 miles) and took a swim.

Answer Soon,

A brother,

Leo

Camp Granite, California
Co. F 359th Inf., A.P.O.-90
28 September 1943- **Day 545**

Dear Zell and All,

How does this find everyone in New Mexico? I am O.K.

I got a three-day pass and visited Cordia and Leveda on Friday, Saturday, and Sunday. We sure had a lot of fun.

I think I will have a chance to go to the Boulder Dam in Nevada this weekend. I sure hope so. The transportation will be free—I hope.

It hasn't been very hot for the past couple of days. It really gets cool in the night here now. I sleep under a couple of blankets.

How is the chance of crops this year? Will you make a half of a crop or better (I hope).

The bulletin board says we get up at 4 A.M. in the morning and go out on a 24-hour endurance hike of some kind. It also said there would be no breakfast. We will get some canned C-Rations sometime tomorrow.

News is very scarce.

Write soon.

A brother always,

Leo

Camp Granite, California
Co. F 359th Inf., A.P.O.-90
9 October 1943- **Day 556**

Dear Zell and All,

How does this find everyone in New Mexico? How is your crop going to make out? I hope it will be better than you thought it would.

I have just got back off a weeklong problem out near the Colorado River. We eat C and K canned individual rations all week. Boy what I wouldn't give for a home cooked meal!

I have really drawn my belt up at breakfast time. I can't stand that hash and stewed meat and beans. The smell will knock you down.

It rained enough to get us and our blankets wet last night and that was all. It is cool and cloudy now and the wind is blowing.

I was aiming to go to L.A. again this weekend, but couldn't get a pass and I had a ride in a care. I sure did hate to miss it.

Have they got the care tires recapped yet? Maybe they will make it sell better. But, I wouldn't give much for them. I hope we can sell it for a good price, but I don't want to sell it cheap.

My squad leader won $190.00 in a crap game yesterday. He started with $2.50. He sure was lucky. But, you can't tell about that kind of money. He may be broke tomorrow.

News is very scarce indeed. Tell Allen and the kids hello and write me a line.

Love always,

Leo

California Desert
Co. F 359th Inf., A.P.O.-90
13 October1943- **Day 560**

Dear Zell and All,

Hi Allen,

How does this find the grain farmer? I am O.K. as far as I know.

We aren't on maneuvers yet. We have just been having small unit training since we have been here. As far as I know, we will leave here on or close to the 22nd of this month and go to Arizona and start Regiment and Division maneuvers. I think they will last until 11

December and Damned if I know what to expect then.

I have heard a lot of rumors. I don't take any stock in them.

I haven't learned a damned thing and I do mean not one thing since we have been here and I doubt like hell if we will before we leave—but I hope we do.

The war department says if we haven't had a furlough in six months that we will have one before we go across.

Our six months will be up in December, so if we go across, maybe we will get another furlough.

Day before yesterday, we had a shot in the arm. It really made me sick. I was on K.P. yesterday and sweated it all out. I guess I feel all right today.

Have you started combining yet? I hope the grain turns out better than you think. From the way things looked when I was last there, I bet you have as good a crop as there is in that country.

Have you had a bite on our car yet? Those tires might be all right to sell when they are recapped, but I wouldn't want them to run. Would you?

Did Delbert fix or have the brakes fixed on the car? I don't think it would sell very well without any brakes.

I have been to Los Angeles to visit Cordia twice. I really did enjoy both of them, though the last one was very short—on a weekend.

We go out on a Battalion combat firing problem this afternoon and stay four days. Everything will fire from a rifle to 105 cannon including the anti-tank guns.

I don't mind it if we fire real ammunition but to hell with the dry runs!

Write anytime,

Leo

P.S. According to Washington, this Division artillery has the best firing record in the U.S. Army. That takes in a lot of country, but that is what General Marshall says.

Desert Training Center, California
Co. F 359th Inf., A.P.O.-90
16 October 1943- **Day 563**

Dear Zell and All,

How does this find everyone? I am O.K.

We just got back off a 4-day problem. We really fired up a lot of explosive ammunition. We fired the big guns as well as the rifles, machine guns, and mortar.

Next week we will go on a similar mission, only we will have air support. They drop real bombs and strafe the ground with 50 caliber machine guns from the P-39s.

It will be an interesting problem and I expect it will cost plenty of money to put on.

Delbert says he is having a lot of combat tactics training now. I expect he will get a lot more.

One of our Company officers died yesterday. I don't know what was wrong with him. He was a Second Lieutenant. He hadn't been with us over three months. I don't think the boys he was over were very sorry. Of course they had rather gotten rid of him otherwise.

Have you had any more bites on the car? I hope it sells O.K. I don't guess it makes a real difference, but it might bring more before the war is over, but no telling what will happen.

If there should be an emergency sickness or anything, you can telegraph me:

PFC Leo Brown

Co "F" 359 Inf. APO-90
Desert Training Center
Camp Young, California

Dear Zell and All,

Odell Hilburn is being stationed here within ten miles of me but I guess I will be gone by then. We leave the 24th.

By the way, do you remember 3 years ago today? It was the first draft registration.

I think the 85th Division is going across soon and Lehman was trained for a replacement for that Division. Did he something about it? It might be all bull, but I don't think so.

Write when you can,

A brother always,

Leo

Desert Training Center, California
Co. F 359th Inf., A.P.O.-90
30 October 1943- **Day 577**

Dear Zell and All,

How does this find everyone? I guess I am O.K.

Boy that cake and candy really came in good. The boys gathered around and helped me eat it. They all bragged on it. Anything homemade really tastes good when you eat out of a can for a month straight.

How is the combining going now? Better I hope. Are you all done cutting at home yet?

We have two more weeks here in Arizona, yet it doesn't make any difference to me if it is Arizona desert or California desert. They are both alike to me.

Does Glen think he will have to go to the Army any time soon?

Tell him to drain the radiator on our car. I don't want to take a chance that it will freeze up

I damned near froze up on the desert last night and now I need a fan and some ice water.

Have they ever set a price on grain yet?

Thanks again for the chow.

Write soon. I just don't know anything to write.

A brother,

Leo

Camp Granite, California
Co. F 359th Inf., A.P.O.-90
14 November 1943- **Day 592**

Dear Zell and All,

How does this find everyone? I am O.K. I guess.

I really have the blues today—the worst ever I had, I believe.

I phoned Cordia this morning but we didn't have much to say to each other. She says, "What do you think about me?"

I said, "Well we won't have time to talk about that in three minutes."

She was still in bed when I called about 7:30 I guess.

I guess she has written to you that she is going to get married to a soldier out there. I think I wrote her a letter every day since then. I tried my best to go out there this weekend but the station wagon we called never came, so we stayed in camp.

I have written everything I could think of to keep her from marrying this guy, but from the way she talked, it was no use.

Never in my life was I so shocked. I got one letter from her. She said what she had sent Ray for Christmas and all you could hear was "Ray." Then, four days later, here came the letter she was getting married to this Sergeant.

I met with him while I was on my three-day pass, but I don't know much about him. He is a good-looking guy and about 23 years old. He is taller than me and weighs 170 lbs. I don't remember but I think he must have about six years in the Army.

I am sure he had two service stripes and two overseas stripes—I think. I know nothing bad or good about him. Maybe it was just because Ray was such a good friend of mine, but in my book he would stack up a good as any of them, and better than a hell of a lot of them.

I also got acquainted with this guy's brother who is in the Navy. The one in the Navy is married. He is the youngest of the two.

We all went out together while I was out there. The sailor is about as funny a guy as you have ever seen. His wife and he seem to get along fine. Like I said, I wouldn't discourage you or encourage you with this guy Cordia says she is going to marry.

I do know it really is going to be a blow to Ray because he thought no one in the world was like Cordia.

Glen wrote me about Papa wrecking his car. That sure was tough, but I am glad he was not hurt. I am also glad he had it insured.

Zell, I believe about the best thing to do is put our car like it was and cover it up the first time you have time.

You were talking about Glen getting worried about what people said because he wasn't in the Army. You can just tell them when they get two sons in the Army then they will have something to gripe about.

I really hate to see him get in the Army as young as he is.

I saw a show last night here in the desert that I had seen about a year before I got in the Army, "Caught in the Draft." But, it really was funny.

Is everybody about caught up with the work there now?

I guess I have about blown enough for this time.

Write soon,

A brother,

Leo

California Desert
Co. F 359th Inf., A.P.O.-90
13 December 1943- **Day 621**

Dear Zell and All,

How does this find everyone? I am O.K.

We got done with our maneuvers—they were over the 10th. I got your letter and Christmas card that day.

Zell, I think Delbert has already sent you some money to buy you and the folks at home some kind of Christmas present. He said he would send $6.00 for everyone except Mamma and he was sending her one personally. So, I will send $8.00 and you can buy her something and put it from me for $2.00 and put our $12.00 on anything you see fit.

Maybe Delbert has already told you something. If he has, that is O.K. for me too.

I think we will leave here probably in January and I think we will go east. Where, I do not know, but they are sending a quartering party ahead (3 per Company) to the camp we are going, to have it in shape. They always send someone ahead.

I had two letters from Ray lately but he never has mentioned Cordia once since they got married. I just answered it yesterday.

We made it to Needles O.K. We like to have run that Negro Division to death. There was a 2-star General from Guadalcanal and a 3-star General from Alaska. They both commented that the 90th was O.K. and ready to go.

I don't know when we will get November pay, so I will send you a check. I am going to send some money (by money order) to the bank after pay day so I can make it $8.00 more and this way I can get you on the road so you can get the Christmas presents.

That isn't much but maybe you can find something for that—I hope. I can't even buy a Christmas card here.

I am going to get my overcoat and go to the G.I. Movie so, Bye

Write soon.

A brother Always,

Leo

California Desert
Co. F 359th Inf., A.P.O.-90
15 December 1943- **Day 623**

CHRISTMAS CARD

Dear Zell and All,

Well, we got our November pay today.

I can tell you about a half a dozen camps we might go and then when we get there we probably wouldn't go to any one of them that I have heard of, so, I will just wait until we get there.

I Hope everyone has a Merry Christmas and a Happy New Year.

Kay Kiser is playing at Camp Young today. One out of each Company got to go. The boy I run around with all the time drew the lucky number.

Love to all,

Leo

FORT DIX, NEW JERSEY

I We boarded trucks for Needles, California (about 80 miles), then we got on a Pullman train car and headed for Ft. Dix, New Jersey (a long train ride). We crossed New Mexico on January 1, 1944. We knew it would be a rough year, but I am glad we didn't know how rough.

Somewhere in New Mexico, a poker game developed. We had a lot of time and not much money, as we hadn't been paid in two months. I had a few dollars, so I couldn't lose much. We played quite awhile, but not much money changed hands as we were playing for table limits (that meant you played with and for the money you had started with on the table in front of you). Well, Willard Potter was dealing. I got three aces and a joker, which made me have four aces (the jokers were played as aces or in any straight of a flush). When it came my time to bet, I raised a dollar and Potter raised me four dollars. Everyone fell out but me. I called him and only had a five-dollar bill left. He dealt me the other ace making it five aces, which couldn't be beaten. He dealt himself two more cards. I told him that I only had five dollars and I would be in for the pot. He said, "Get out and borrow some. It is OK with me."

I told him, "I can borrow money on it!"

He thought I drew two pairs and was bluffing.

One fellow said, "I haven't seen either hand, but I will let Brown have a loan if he wants it."

I told him, "Thanks, but I will go on the rules we started."

Potter was still trying to get me to borrow some money. I was sure he had a good hand, but he was from my platoon and a friend, so I put in my five dollars as he did, and I turned my hand over, and he threw the rest of the deck of cards all over the train car! So, that broke up the poker game. He only had three kings and thought that I only had two pairs. I only came out with twenty dollars.

We pulled into Albuquerque, New Mexico about noon, and it was announced how long we would be there. There was another New Mexican in our platoon besides me, Frank Savarda. He called his wife, and she came and visited with him. The next morning, just as the sun was coming up, we pulled into Dodge City, Kansas. I can still remember how friendly the people there were and how cold the weather was. They gave us cigarettes, coffee, and cupcakes.

I remember how beautiful the farmsteads were in Iowa and Illinois. Then, I remember getting in at Chicago about midnight. We had to change trains and train stations. I think it was about twelve miles. We had to carry everything we had and ride open trucks. It must have been zero degrees, the wind fifty miles per hour, and sleet that felt like broken glass. We caught the train and continued east. We went by way of New York City on our way to Trenton, New Jersey. We went to Fort

Dix. It was a real fort with two story barracks. Other than the damp cold weather, it was the best camp we had ever been in. I was surprised to learn they were still heating and cooking with coal.

There were a lot of guys who came down with the flu. Fifty percent went on furlough and the others who were sick tended to the fort and stood guard.

I learned that what I didn't like about Fort Dix was it was "wet" in camp. To me, that encouraged too much drinking, the same ones trying to drink all the beer up each night. Fort Dix had been there a long time. They had good theaters, a big field house where they had USO-sponsored dances, and lots of activities, quite different from the desert. To my way of thinking, the desert was better. There were too many bright lights for a country boy. Well, they were giving furloughs, and we still hadn't been paid. I had about ten dollars, and a train ticket home was about fifty-five dollars. Some of the men were getting a loan from the Red Cross. I stood in a line for about two hours and finally got up to the desk and stated my business and I said that I could send a money order for sixty-five dollars as soon as I got home.

They asked if I was married, and I said that I wasn't. They said that as a single man, I wouldn't be loaned money at all. Most of them got an old gal and shacked up and didn't go home anyway. I won't tell what I said to the man at the Red Cross other than that I told him what he could do with it!

I found out that you could get a check for fifty dollars from the post bank if you had an I.D. card, which I had, so I wrote a check for fifty dollars. I went out and walked around a while, had a Coke, and then wrote another check for fifty dollars. Then I went to another teller and wrote another check. Well, I was fixed up for money. That night, there was a guy from our company, not our platoon. I knew him and who he was, but I had never talked to him. He came down (he lived in the top story of the barracks) and came to me and said, "I heard you were getting a furlough and needed money. I have plenty of money and no furlough." I figured this would be one of those high-interest loans. But, he said, "There won't be any interest!"

I couldn't figure it out. I said, "There are probably a lot of guys in your platoon who would love to have a loan."

He said, "But I don't think they will pay me back, and I know you will."

His comment touched me, though I didn't know why he was aware of that.

He was right. I knew a guy from Arkansas that couldn't get money, so I borrowed on his behalf. As soon as I got home, I sent him a money order and waited until payday for mine. I never did forget that.

I went to get my ticket, and they wanted to route me through New York and Chicago. But, I told them that I wanted to go through Pittsburgh, St. Louis, and Kansas City. Well, we were to catch the train from Philadelphia. There was a guy from our company with me, and we were waiting for the train. There was this black man, a white woman, and a black baby. We hadn't seen that out west yet. The guy I was with was a fist-fighter and troublemaker. He said, "Brown, let's kill that 'nigger'!"

I said, "She made that choice, and I am going home on furlough. You do what you want, but you will land in jail!"

He took my advice, and we caught the train. It was very crowded. We had to stand up until we got to Pittsburgh. There, they hooked on more cars. Boy, I was tired! So, I leaned back and went to sleep. When I woke, it was warm (hot water pipes heated the car). I noticed my feet were really warm. I looked down and saw hot water two inches deep in the floor. The car I was in had been out of service a long time.

Somewhere along the line, the Salvation Army gave us a cup of coffee and a cupcake. That was all we had until we got to St. Louis. It was midnight and the military police (M.P.) didn't want us out of the station but finally let us get something to eat, a half chicken each. The next stop was Kansas City. We left there, and it was the same old story. I still had nowhere to sit and when the porter came by I said, "I have half of a dollar for a seat!"

He said, "Hang on to it, and I will be right back."

Sure enough, in a little while he came back and took my half-dollar and got a slot (I guess he threw someone off). I don't know how he found the seat.

We arrived at Clovis, New Mexico, about 2300 and discovered that there was no train to Portales until 6:00 A.M. So, I walked to the bus station, which was some walking. It was closed up and dark. I started back to the train station, and an M.P. picked me up and took me to the depot and told me that if I got out before morning, I would be locked up. I waited until 6:00 A.M.

I went home to say my goodbyes and fix up my financial affairs, what little I did have to fix up. My brother Delbert was on furlough too. When there was an opening for furlough, we would each send the other a telegram stating when we would get the furlough. We would go to the company commander and show him the telegram and tell him how much it would mean to have our furloughs together. It worked. We did it about three times.

Delbert was dating a girl out at the college. She had a good-looking roommate I dated several times. She wouldn't admit it, but I think she believed in fortune tellers. She went to three while I was gone and they all told her, "That guy is not coming home!"

We stayed in touch by letter while I was gone. I was glad she didn't say what the fortuneteller told her. I think she was a little surprised when I came home.

Delbert and I left together. For some reason, we left on the bus to Amarillo. There, we got on different trains. It was a real cold day. There weren't many on the bus. There was a girl laying on one of the seats trying to sleep but was too cold. Delbert asked her if she wanted to borrow his GI overcoat. She did. When we got to Canyon, Texas, the military police got on the bus to check furloughs. Delbert and I were the only servicemen on the bus. The M.P. thought the girl under the overcoat was a soldier and he began to yell at her to get up and show her pass. Nothing happened, so he jerked the overcoat off, not very gently. She called him some names that I hadn't heard before, and I had been in the army for nearly two years.

We got on the train to Amarillo. The next day, we arrived at Kansas City. It was a big terminal with lots of people waiting to get on the train. I checked and found which door we would be loading at and got in line close to the front. Before the train was loaded, someone came over and said that servicemen would be loading first. So, we started a new line, which I got in.

A drunken sailor had his white sailor bag and got in the back of the line and said, "Men, we will line up! In the front will be army and navy cadets."

He named off several other ranks. Then, he said, "In the rear will be soldiers, sailors, and dogs."

We got on the train and headed out to St. Louis. The drunken sailor stood up in his seat and said, "Men, I am going to meet my wife in St. Louis and I want to be good and sober when I get there. So, I have this pint of whiskey to sell to the highest bidder."

Some sergeant offered him twelve dollars for it, and the sailor said, "If it is worth that, it is worth too much to sell!"

So he sat down with the sergeant and they drank the whiskey. Sure enough, his wife met him at the station, and he had to be helped off to her.

Nothing else of interest happened and I got back to Ft. Dix about three hours early. Then, I started worrying about my good buddy, Sullivan. I always dreaded him leaving camp for fear he wouldn't come back. I wrote earlier about him getting a three-day extension to see his brother in the Navy (he didn't have a brother). Things like that were a real challenge to him. With his extension, he got back on time!

Ft. Dix was called a staging area. We checked and rechecked our equipment and clothing. I think mostly to have something to do. That was a lot better than the cold outside. This fort was old and had lots of extras. There was a big basketball gym. We went out and played basketball. We didn't have to do the marching (except parades and exercises). So I went for it! One night there was roller-skating at the field house. I had done a little roller-skating in my time, so I went. There was music, and you had to have a partner. The U.S.O. brought lots of girls from New York and other places close by. I was standing and watching when this girl came up and asked me, "Do you want to skate?"

I told her, "I have never skated except by myself and never to music."

She said, "If you can march to a band, you can skate to music."

It worked out O.K.

She said, "The U.S.O. Is going to bring the girls next Saturday night and have a dance." She asked, "Could I depend on you being there to be my partner?"

I told her, "O.K."

I think there was a lot of prestige in coming to the dance with a partner instead of as a group. So, the next Saturday night she was there, I told her, "There are two things that I want to do. One is to go to New York City, and the other is to go the Washington D.C."

She said, "I would be delighted to show you New York. Why don't you get a buddy and come to our house, spend the weekend, and we will go see New York City?"

I said, "Let's hold on a minute! What will your father think of a couple of soldiers coming for the weekend?"

She said, "I'm sure it will be O.K."

She gave me the address and phone number. It was in East Orange, New Jersey, just across the river from New York. Well, I got Sullivan, and he was sure that we would never find it. We got on what we hoped was the right bus, and when we got to the place where we thought we should be, I asked the bus driver about it. He said, "Boy if you will sit down, I will let you off at the right place."

He let us off, and we found the right place.

Her parents were friendly to us, and in a few minutes, we had a real good meal. The girl went over and talked to her dad. He had a schedule made out and what buses to ride and how long each place should take; we were to be home by 11:00 P.M. They said there was one place to go, and that was to see the movie at Radio City, a seventy-story building. Sometime in the evening, we got there, and Odell Sullivan and I said that it was a pretty good show, but that we had seen it. They said, "No, that's impossible! This is the very first place they show when they are produced on Broadway."

That defeated them because it was supposed to be a feather in your hat to see the movie at Radio City. We had seen it at the army base for 15 cents. We didn't see that show, but saw another one instead. The next morning, we got up about 7:00 A.M., and a few minutes later, we had breakfast. Then her dad had a talk with her and gave her a schedule. We were to be back by 2:00 P.M. for dinner. We had a look at New York in daytime and sure enough we got to the house at 2:00 P.M. and ate and visited a while. Odell and I thanked them so much and said, "We suppose it is time we got back to camp."

The daddy asked what time we were to be back at camp and we said at 11:00 P.M. He made out another schedule and said, "That will put you at Ft. Dix and the girls will be within two blocks of home."

The whole family was well organized, but I never saw anything like him. He said, "We try to do this once a month. It makes us feel like we are doing it for the war effort." I don't know, but that kind of left a sour taste in my mouth! If he hadn't said anything, I would have said that they were very nice. But, when it became a duty, it changed it some. He didn't say, "Come again" or "glad to have you by," it was all over, and we had our share. Like I said, what he said changed things.

Well, in another week or two, we got a weekend pass and were going to Washington D.C. We got to Philadelphia and went to the bus stop. People stood in a long line that went back and forth across the big parking lot.

We decided we would never get there. There were a couple of girls in front of us. They said, "We will show you Philadelphia."

It was cold and snowing hard, so we picked up the girls and gave up Washington D.C. Really, we figured they wanted to show us the nightspots and bars. Pretty soon, I told them, "We have had enough for one evening, and besides, we are about to freeze to death!"

They wanted us to escort them home because it was dangerous for them to be out. They told us which subway to get on. It wasn't far from their home, and we left and boarded the subway with instructions for where to get off.

A marine and his sister were on one end of the car and about fifteen black "Zoot Suiter" gang members. It looked like big trouble. It was plain that they were drunk. One of them said, "I want to kill somebody about like those three soldiers back there (meaning us)."

We put the girl back in the corner behind us, and the marine whipped a knife out of the top of his boot. It was about a 10-inch blade. I had a big pocketknife, but not like that, and so did Sullivan. The marine said, "Come on back, I've always wanted to gut a Zoot Suiter like you for a long time, and I may as well start with you!"

We held them at bay until the next stop. We got off. They didn't follow.

When we were sure they weren't going to follow us, I asked the Marine. "Were you afraid?"

He said, "Hell man, I was never so scared in my life! I was never so glad to see two guys as I was you two!"

I said, "We could have taken a few, but the outcome didn't look so good."

He said, "Yes, but none of them wanted to be first."

We called a cab to get us back to camp, and I haven't been on a subway since.

Long about this time I got my delayed birth certificate, but they wouldn't talk to me about trying to be a "fly boy."

Somewhere in the area, the army was breaking up the regiment of a division that had four regiments instead of three. We got a lot of replacements that we didn't need. A lot of them were sergeants—they were older guys of the regular army. Usually, those were the ones you wanted to stay away from if you could. You can figure, that if an outfit is cutting its numbers, they aren't going to ship out the good guys, it's the wet ones you get. There was quite a lot of friction between the old regular army men and the draftees, who outnumbered the regulars by ten to one. The draftees were young, bright, and made a lot better soldier.

1944

Fort Dix, New Jersey
Co. F 359th, 90th Infantry Division
26 January 1944- **Day 665**

Dear Zell and All,

How does this find everyone? I am O.K.

I got back to camp a day early and really made good connections coming back. I had the streamline train all the way from Kansas City. I just hopped off one on to the other at St. Louis. It cost quite a lot to eat coming back. I had to eat on the diner all the time. It cost $1.50 per meal and I had to buy apples, candy, sandwiches, milk, etc. to get by.

Delbert and I rode the bus as far as Amarillo together, ate a good steak dinner there, then shot a couple of games of pool, then took off.

Zell, we had our proofs taken at the photo shop by the Y&M Theater and paid $1.00 on them. You can go in and look at them if they are O.K. We want you get us three of the $1.25 ones for Momma, Iva, and Cordia. Mail Iva's and Cordia's to them then send and tell us how much each of us owe you and we will send the doe.

You an go ahead and get a big one for $2.50 or take a $1.25 and figure the difference when you send the bill. I am sending a buck of your money. I paid one dollar on them. I don't know if that will come off the pictures or not, you can tell us. I would have sent enough money but it is a scarce article. Now, I guess we will get paid the 10th.

Has Glen brought the car over there yet?

I saw a show last night, "Fighting Seabees."

I believe Dale would think the Seabees would be a little rough.

News is scarce. Write soon.

Love always,

Leo

Fort Dix, New Jersey
Co. F 359th Inf., A.P.O.-90
3 February 1944- **Day 673**

Dear Zell and All,

I got your letter today. I guess you have received mine with the pictures before now ,or at least you should have them.

Were any of the proofs any good? If you haven't received the letter with the dollar bill with the instructions I included, let me know

Maybe Delbert has told you differently but we want 4 of the $1.50 ones: One for you, Iva, Mamma, and Cordia. He said he didn't want any more in uniform and I don't either..

You can pay for it and tell us. We will split it and pay for it. Mail the ones for Cordia and Iva then put it all on one bill. There is not use sending it to us and us sending it to them.

I paid a dollar for them and put a dollar in the letter that I wrote you a few days ago. Let me know when you get it.

Things are going about as usual here—the same old grind.

I have played a lot of basketball since I got back. It sure made me sore for the first three days.

I had good luck coming back. It has been plenty cold the last few days.

Write soon,

A brother Always,

Leo

Fort Dix, New Jersey
Co. F 359th Inf., A.P.O.-90
6 March 1944- **Day 705**

Dear Zell and All,

How does this find everyone? I am O.K. and feeling fine.

We have had a lot of rain and snow this week, but the sun rose clear for the second time in two months this morning, but it didn't stay clear. It looks like it might rain any time now.

Zell from the way you wrote, I gave you the wrong opinion about New York. It is a big city and all that but I would take New Mexico and hher sand storms any day and you would too.

I could write three or four pages on the reason why, but when I got through, it would all amount to one thing—that is home. We understand the ways of life there and the people. Here, everything is much different.

I know there isn't much news in the letters I write up here like the Captain said about the weather or what we had for dinner, but we can't write much. If you don't hear from me for a longer period of time than usual, don't think anything about it because I will be O.K. Do not say anything about it outside of the family. There are too many people interested in what Divisions are on the coast are doing.

This isn't a way of letting you know we are leaving or going across. I don't know anything of importance myself but in the future I will probably more than I can write.

So, for the time being, I will write about the weather or what movie I saw and like. By the way, there is a good one I intend to see tonight "Purple Hear" here in a camp theater.

Did you have any luck selling the Plymoth? What did Delbert say about it?

Answer soon,

A Brother Always,

With Love,

Leo

P.S. Tell Don he can have that little khaki cap that you said fit him. It wasn't mine. A guy was going to throw it away and he asked me if I wanted it.

Fort Dix, New Jersey
Co. F 359th Inf., A.P.O.-90
23 February 1944- **Day 693**

Dear Zell and All,

How does this find everyone? I am O.K.

I sent my watch home yesterday and a little surplus junk today. You can take it to our home some time—there is no hurry. Mamma might think it meant something but it doesn't mean anything. We, as you know, have don the same thing over a year ago. The watch for some reason stopped.

I went to New York last weekend. I saw a lot of sites. New York is a beautiful city—the most beautiful I have seen, but it is much too big for a country lad.

I went with a girl that lives 20 miles this side of there. She knew the town pretty well.

We went down Broadway, Times Square, 5th Avenue, and a lot of places I don't remember. The most interesting thing I saw was the Rockefeller building. We went on top of it, 70 stories. You could see the entire town from there. It is about 2 miles from the Empire State Building. It was easy to see and you could see the Chrysler building, Statue of Liberty, Wall Street, the Hudson River, the East River, Long Island, and the Atlantic Ocean.

It takes 15,000 people to occupy the building. There were shows, an ice skating rink, and all kinds of stores to buy anything from a spool of thread to a battle ship.

It is entirely supplied by the subway. Radio City is in the building, is the largest theater in the world, and seats 6,000 people. Inside is a 70 foot screen and a lobby that is 70 feet high. The lights weigh 3 tons each. There are businesses for several foreign nations and many interesting sculptures.

We gave guide 55 cents each to take about a dozen of us through and tell us about it. It also has the fastest elevator in the world. It makes the first 65 floors in 30 seconds. It gave us a funny feeling to ride it.

I went under the Hudson River in a subway three times and by way of the Staten Island Ferry the other time.

Guess you are tired of this bull.

Write soon.

A brother,

Leo

Fort Dix, New Jersey
Co. F 359th Inf., A.P.O.-90

25 February 1944- **Day 695**

Dear Zell,

I just received your card. Yes as far as I am concerned you can sell the care for $650.00 if it is O.K. with Delbert. It is O.K. by me if you can get $600.00.

I saw a really good show last night: Wallace Berry in "Rationing." It was really funny. I saw a good basketball game Wednesday night. Fort Dix and Fort Hamilton played and the score was 41 to 39 in favor of Fort Dix.

The Fort Hamilton W.A.C.s won over Fort Dix's W.A.C.s by 20 to 14.

I sure would like to go to Washington D.C. this weekend but I am broke and Tuesday is payday.

Have received my watch I sent?

The sun rose clean as a whistle this morning, like it does there at home—the first time since I have been there.

I hope the guy takes the car for $650.00

A brother,

Leo

"Things were winding down at Fort Dix. We were sending things home, making allotments, and having pictures made. We had a photo made of F Company 359. I have often wondered how many of the old Arizona F. Company came home. I bet not many.

We moved to Camp Kilmer, New Jersey. It was a good camp but not like Fort Dix. Kilmer was the port of debarkation, not far from New York Harbor. We were there two weeks; then we got the word that we were leaving. We just had two more nights if we wanted to go to New York. My total bankroll was $2.00. Even in those days, you didn't go too far with two bucks! There were several in the same shape that I was in, so, we started a friendly little dice game. By bedtime, I had won $30.00, but it was too late to go anywhere. I was sure I would go the next night, but I caught K.P. The night after that, we shipped out.

THE CROSSING
PREPARING FOR INVASION

I We left Camp Kilmer and loaded on the ship in the dark with a cold rain (nighttime was for security reasons). We were on the deck of our ship when daylight came. Through the rain and fog, we could see the Statue of Liberty facing out to sea. We were told to go below deck so that no one could tell it was a troop ship. Ralph Duncan said, "

If I ever get back, that old gal (the Statue of Liberty) will have to do an about-face to ever see me again."

We were assigned bunks that were stacked four high. Everything was full. We had chow. It wasn't too good. I don't remember the name of the ship, but it was a medium-sized British ship. After we had gotten out of the harbor, we could go up on the deck anytime unless the weather was really bad. Ralph Duncan went up on deck as soon as he got up, and he came back down and said, "Boys, we didn't move any last night!"

I asked," How do you know?"

He said, "The scenery looks the same; nothing has changed!"

And it didn't change for fourteen more days. We were in a big convoy. We had been out to sea for three or four days, and it was Sunday. The water was calm. We had church services up on deck. I thought if all sailing were like this, it would be O.K. The weather was warm, but sometimes, at night, we could feel the ship bucking and rolling.

One morning, the waves were coming up on deck. We were ordered not to go outside. We might get washed overboard. About breakfast time, the captain of the ship said, "Boys, we have got it made now! No plane or sub can operate in this weather. Yesterday was the worst I have seen."

We hadn't even thought about it. With the rough seas, the guys began to get seasick. I didn't feel well, but I wasn't vomiting. The first sergeant was blowing off and telling the men that they were like a bunch of kids. He was trying to make it as tough on everybody as he could. Then, I noticed he began to look a little pale. One night, I woke up, and the ship was bucking. I could see by nightlight that it had thrown the first sergeant out on the floor. He got a bucket that he hid and said, "Oh my God! I believe I am going to die!"

He was sure everyone was asleep. I giggled just enough to let him know I heard him and thought it was funny.

Someone from our platoon got a little poker game started. None of the others were poker players, including me. I still had the $32.00 I left with, so I got in the game, played a couple of hours, and won a little but not much. Then someone dealt me an ace high straight. So, I bet pretty well, and everyone stayed in. I said, "I don't want any. I will play these." Everyone drew three cards except one person. He drew four cards. Well, I bet all I had and everyone called, and there were a few side bets. I had the lowest hand of the six players.

One had a flush and one a full house and one had four deuces. That ended my poker playing because I realized if I were in a big game, I probably would have lost everything. I was so sure I had a winning hand. I didn't even come close. I lost my money, but I learned a lot more. I was one guy who landed in England that didn't have to change any money.

There were a lot of rumors as to where we were going, North Africa or Italy. The English sailors said all the time that we were going to England, and they were right. One of our boys who was seasick asked one of the English sailors what was good to stop seasickness. The sailor said, "Well, ol' boy, the only thing I know is to get your two feet on solid ground."

At the time, that was two miles—straight down.

I think we stayed close to the coast of England as we went north and into the port and unloaded at Liverpool. There was a lot of bomb damage there. The Germans tried hard to sink enough ships in its harbor to close the port, but those Englishmen just wouldn't give up and kept it open. We had seen a lot of movies and training films, but this was the first real damage from the enemy that we had seen. We were told that there had been so many ships sunk around England that oil was twelve inches thick around the beaches. Boy, what the environmentalists would do with that these days!

I think this was March 22, 1944. I do know that it was Friday or Saturday. The following Sunday was Easter. A lot of us rode GI trucks to what we thought was a Protestant church, but I think it was the Church of England. The service was in Latin. We didn't understand a word of it, but we understood when they passed the collection plate, and we all put in a pound note. It made the papers! The church was over a thousand years old, and that was their best collection ever.

They located our battalion near Birmingham. Our tents were scattered out among the trees for concealment so that not too many would be in one place in case of an air attack. We were in tents with no floors or lights. The latrines had no running water. We shared a shower with another unit. Laundry was a real problem. It rained so much, it was hard to get anything dry. Things were very crowded with other servicemen from the U.S., Canada, and just about everywhere else except for enemy countries.

Sometimes, Sullivan and I would go to town. You could always get fish and chips and a pint of bitter (fish, French fries, and a pint of bitter beer). The British were good people and treated us better than we deserved. There were a lot of British women that I couldn't say that much for some of them were hanging on to a "Yank" (an American), and the English servicemen were alone or usually with another English soldier.

England was very glad that we had entered the war on its side. If there were any doubt about the outcome, they never showed it. We would have had a tough time making a landing on the western front alone. First off, the British and its entire commonwealth had a great reserve of trained reserves. Most of them were combat tested, for which there is no substitute.

The British and Americans had enormous stockpiles of material hidden under groves of trees that included thousands of 55-gallon barrels of fuel, new trucks, and ammunition. We had little room to train. There was mostly road marching, and everyone had a full load of ammunition, which added a lot of weight to what we were already carrying.

My record showed that I had done some work as a mechanic. So, I went to school for a week to learn how to waterproof vehicles so they could drive them off of a barge and not sink and drown. I went back to the company. The British were like us. They had a lot of time on their hands, and they got the trucks fixed early, so they never called me.

Ten days before D-Day (June 6, 1944), we moved to a camp in Wales. They enclosed our camp with wire, just like a prison camp. We in the Second Battalion shared this camp with the 101st Airborne. They had more men than they had planes and gliders.

We got one pass to go into town. All the pass said was, "John Doe has permission to be absent from camp until 1100 hours."

It didn't say which camp or unit from which we would not be present, but we knew that time was growing short. Odell Sullivan and I got a pass and went to a town in Wales. We didn't stay very long. They told us that we'd better hang on to the pass because there wasn't any way to get back into camp. We got back early, and when we got to the gate, the guard took my pass and let me in. Odell said, "I am going to try to get in without a pass."

I knew Odell. He put his hands in his pockets and said, "I guess I lost my pass."

The guard said, "I hope you haven't, because you can't get in without it."

Odell said, "You remember me? I was with Brown here, and he had a pass."

The guard said, "There have been about 500 men going in and out of here, and I can't remember each one."

There was a soldier who had gotten in and still had his pass. He got up close and dropped his pass next to Odell's feet.

Odell said, "There it is! It had just fallen out of my pocket."

The guard saw it. People were piling up on each side of the gate, wanting in and out, and the guard, feeling the pressure, said, "I suppose you were with this guy [me], and he has a pass, and you somehow got out, so I'll let you in."

As he walked in, Odell handed him the pass and said, "I wouldn't want you to get into any trouble."

Somewhere In England

Co. F 359th Inf., A.P.O.-90
7 April 1944- **Day 737**

Dear Zell and All,

How does this find you and everyone? I am safely in dear old England. The place where we are now is nice and Spring is nearly here now.

The trip over was O.K. The food was poor and we were crowded, otherwise it wasn't bad. Time went fairly fast. I was shade seasick for 24 hours but never long over the rail and fed the fish any.

Did you have any luck selling the Plymouth?

I haven't gone any place yet. We turned in our money today. I should get about 9 pounds and 4 shillings.

Bye for now,

Leo

Note to the reader. According to Wikipedia the "V Mail" system was developed by the British in order to save shipping the original letter across the ocean.

"V-mail, short for Victory Mail, was a hybrid mail process used by America during the Second World War as the primary and secure method to correspond with soldiers stationed abroad. To reduce the cost of transferring an original letter through the military postal system, a V-mail letter would be censored, copied to film, and printed back to paper upon arrival at its destination. The V-mail process is based on the earlier British Airgraph." (Wikipedia)

England
Co. F 359th Inf., A.P.O.-90
23 April 1944- **Day 753**
Dear Zell and All,

How does this find everyone? I am O.K and feeling swell.

The Army is the same old thing here as ever. Sometimes the training is rough and sometimes it is easy. There are a lot of Generals drifting around lately.

All the Sergeants and officers got to hear **(CENSORED**). I would like to have heard him talk. I heard another one talk. They say he made the best talk. HE really seemed to be a sensible man. He fought the Japs for over a year.

There is a poker game here in the tent. The boys seem to be having trouble counting their money. I can count it fairly well but I still think the U.S. money is the simplest. We have a lot less coins than they have.

Love to all.

Leo

P.S. I received your letter yesterday.

Note to reader: The censored data corresponds to Leo's memoirs and the person speaking that day was General Omar Bradley.

England
Co. F 359th Inf., A.P.O.-90
8 May 1944- **Day 768**

Dear Zell and All,

How does this find everyone? I am O.K.

Things are going as usual here. The Army way and the training is rough.

I had a letter from the folks yesterday. They had just heard from me for the first time. It sure did take a long time for my letters to get across for some reason.

Has it rained any more there yet so you can plant a crop yet? It is a shade early yet. I hope by next year I will have something like that to worry about, a year ought to change a lot of things.

Love to all,

Leo

Somewhere In England
Co. F 359th Inf., A.P.O.-90
10 May1944- **Day 770**

Dear Zell and All,

I don't remember if I wrote you last or not.

Everything is going along O.K. here—the same old thing.

I went to town last night and went to the show. It was a universal picture, nearly all of them are, so they are O.K. for us Yanks.

Have you had any rain this spring? I hope so. We could spare you a little.

It is O.K. to use V-Mail but it is not necessary, if you can think of more to write. They can hold about what I write.

Love to all,

Leo

United Kingdom

Co. F 359th Inf., A.P.O.-90
21 May 1944- Day 781

Dear Zell and All,

I received your V-letter of May 6th. I got an Air Mail letter from the folks dated 12 May at the same time. The reason I use V-Mail is it is all I have.

Don't ever let anyone tell you that the U.S. isn't a city block ahead of this country. And for food, they don't know what a steak is. I haven't seen one since I left the U.S.—or white bread either.

They really go for the hot tea. They won't run short on my account, as you know.

This is pretty country, but they don't know what summer means. It is as cold now as when we arrived. It is easy to see why we turned in our khaki uniforms before we left the States.

You asked if you could send a package. It is really O.K. by me. It will take about a month to get here.

Love to all,

Leo

United Kingdom
25 May 1944- **Day 785**

Dear Allen,

I received your V-letter of 8 May a couple of days ago. I was glad to hear from you and that you are O.K. and feeling fine.

Have you had to take your physical exam yet? I sure hope you don't have to go to the Army. If you do, I don't think you will have to worry about coming this way. By the time you were trained, I think the war over here will be done. I hope the whole damn thing is over.

Well Allen, I guess this is a pretty fair country. The weather is lousy nearly all the time. Like I said, it is pretty fair country, but I think after the war there will be plenty of people here to run it O.K. without us Yanks. The U.S. is too far ahead of this country to compare the difference.

I thought the British would be nearly as modern as the States. Like I said, they are nothing to compare, however, this country has been at war longer than we have and that makes a difference.

There is no place to go worth a cuss. They just simply don't have anything for amusement and it is impossible to buy a meal. We have to be off the street by 22:30.

On a clear night, it isn't quite as dark. So, to my notion--a G.I. movie is the best bet.

There are some pretty fair looking dames here and easy to pick up, but most of them would take a negro if he has a uniform on just as quick, so you can figure how you would like that.

The Negro and the White soldiers are a big problem over here. The Negros go in gangs and beat up a few white soldiers. They get plenty rough and always have knives and clubs.

If I were you, I would try to get my age straightened out with the draft board. It will be a shade bad if you have to go to the Army. I don't believe Don is quite old enough to take over. Ha.

Write when you have time.

Leo

"We had several inspections. After that, we went to the port. We loaded onto the transport ship, the Susan B. Anthony. It was a good ship, and it was American-owned. This was about June 3, 1944. We sailed from Wales. It was quite a few miles to Normandy, France, from there. As soon as we got out of port, we were called to the dining room and told that were to land D+1, June 7, 1944 (one day after the first day of the Normandy invasion). We were given all the details. There were no more rumors.

INVASION!

Before we went to the harbor to load the ship (there were a lot of other units loading there as well), there were many big generals who came to our camp. Several of them gave us pep talks. General Eisenhower sent a representative, but General Omar Bradley (four stars) gave us a pep talk and told us how lucky we were to have a front seat to the biggest event in history! He said that we would be in more danger at home driving our cars. He told us that when America landed in North Africa that only seven men per thousand were lost. They didn't expect this one to be any worse. I expected it to be worse than that but wasn't sure how much worse. I lost some respect for General Bradley for telling us that.

As far as I know, we traveled all night, and they threw out a few depth charges. I suppose they heard signs of submarines. We heard rumors that we had hit a mine. Later, we found out how wrong we were.

We had breakfast, and anyone who wanted to go up the deck went there for a while. It was just getting daylight, and the coast of France was in sight. There were some fires burning, and there were more ships than I ever knew existed.

There were little ones all the way to the old Texas battleship, which was to our left front and sitting broadside so it could fire all guns, and they were all being fired! We were told to go back to our quarters and wait for instructions from the P.A. system. We started moving in a sea-lane. We were the flagship (the first one). About that time, we hit a mine!

The captain of the ship behind us said it blew part of our ship so far out of the water, he could see daylight under it. I hadn't heard anything like it! The lights went out, the PA system went off, and the steel stairway came loose and flew across the room. I thought about this. There were massive wooden beams across the top with waterproof double tarps over them. I got up on a row of bunks; there was so much confusion on the floor that there was a good chance of getting trampled. I stayed on the bunks and went to the tarp.

The explosion had broken the wooden timber, but I got my knife and cut the tarp, and when I did, a couple of barrels of saltwater hit me in the face. I thought we were sinking, but by then, it was gone. I climbed out on the deck, and many men followed me out that way. Shortly, a British destroyer pulled up on our right (starboard). Everyone rushed to the starboard side to get off. Our ship began to list, and we came very near to capsizing! The captain was up on the bridge with a bullhorn. He wasn't ordering us back over; he was begging us to help balance the sinking ship.

There must have been a terrible fire because three smokestacks had black smoke boiling out of them. I looked up at the sky and saw three P-38 American planes circling overhead. They looked superb! We were sinking pretty fast. Shortly an American destroyer pulled up on the port (left). While we were waiting for the new ship, I noticed a soldier eating an apple.

I asked him, "Are you hungry?"

He replied, "No, but I don't want to get salt water on my apple."

They threw three cargo nets across, and we went across like rats. I was on the British ship. One of the British soldiers remarked how tired we looked. I thought about that and realized that I was exhausted. The British carried buckets of hot tea up on deck. I had never had hot tea before. It was good!

We went for a while, and they transported us to landing crafts. These crafts were flat-bottomed, so they were rough. Some of the guys were sick by the time we got to the shore. The division commander, a one-star general, and a full colonel were on the landing craft with me. We flagged a small Navy ship, and the general asked the sailor if he knew where Utah Beach was. He said, "I don't know, and I don't give a shit!"

The general reminded him, "You are speaking to an officer!"

He replied, "General, your rank won't get you anything out here."

Then, he took off. We did find Utah Beach. We got off the craft in about four feet of water. The beach was relatively clean except for life belts and other junk that had been discarded. There were a couple of German fighters that had strafed the beach, but some Navy guns had brought them down.

We walked inland for about two miles. There were dead Germans along the road. I later found out that we needed to worry about the live ones. The general stopped us in an apple orchard. In the corner was a small triangular depression. The general and colonel got in it and told the rest of us (we numbered about 60), "Dig both an officer and enlisted men latrine and then set up a four-man walking guard."

We chose the other low area, which was also triangular in shape. We had only four rifles in all our armament.

The general never made any effort to locate his command. They just sat there. A sniper started taking pot shots at us. I took one of the rifles, borrowed a pair of field glasses, and put in about three hours trying to locate the sniper, without any luck. There, I realized that my training was lacking. I should have been trained to locate rifle shots.

Everything seemed to indicate that the landing at Utah Beach was successful. I didn't see Omaha Beach. There, the soldiers had a terrible and costly landing. Two regiments of the 90th and 4th Infantry Division preceded us, and we just sat there the rest of the day. By night, those green apples began to taste pretty good.

Well, night came, and we had no food or blankets. We young ones had dried out from being soaked in seawater by now. We made out O.K. but were not comfortable. The general and the colonel, we figured to be in their 60s, were still wet because they sat on their butts all day.

They were shaking so bad, they could hardly talk. At about 10:00 a.m., someone found us, and they sent for some arms and ammunition. When they arrived, we learned that they were mostly German rifles and a bunch of ammo. Most of the ammunition didn't fit the rifles, although it looked the same. It appeared that the general was going to form a military unit out of us with German weapons.

No one knew me or who I was, so I leaned my German rifle against a tree and, when no one was watching, I took off on my own. I didn't think I could do worse than that. I didn't go very far before I came to a first aid station. Up walked a paratrooper with a wounded leg. He gave me his rifle, ammunition belt, canteen, and pack. I saw some wrecked gliders, and in them, I found a helmet and some more stuff, including rations that I badly needed. I took a parachute and rolled it up for a blanket. I found two 60-mm mortars and a bunch of 60-mm ammunition. I couldn't carry them, so I hid them.

Pretty soon, I started finding 90th Division men. Later, I found part of our company. I expected to see Captain Lepene. He had been with us from day one. Someone said his knee had been hurt when the ship hit the mine. My lieutenant was glad about the 60-mm mortars and ammunition that I had found. We went and got them. Things were not the same. A lot of our men were gone, along with our captain and First Lieutenant Raper, the exec. He was a good and able officer. The next day we were given orders to attack. As I remember, we were to attack to the west toward Cherbourg.

I have been dreading getting to this part. I readily admit that I don't know how or why you fight wars. So, I will write a few things that happened along the way, and I will leave the wherefore and why to the generals and politicians. I don't claim this to be an accurate account of the things that happened in Western Europe, but through the eyes of one old infantry soldier, it still saddens me to think of those two days.

Susan B Anthony

"Early in the morning of 7 June 1944, while cruising through a swept channel off Normandy, Susan B. Anthony struck a mine that exploded under her number 4 hold. Immediately she lost all power, and her rudder went hard left and stuck. By 0805 hrs, holds numbers 4 and 5 were shipping water badly, and the ship took on an 8° list to starboard. In an effort to save his ship, the commanding officer, Commander TL Gray, USNR, ordered the embarked soldiers to move to the port side. This human ballast soon brought Anthony back to an even keel.

At 0822 hrs fleet tug USS Pinto came alongside, prepared to tow the paralyzed Anthony to shallow water. However, fires soon erupted in her engine and fire rooms, and she began to settle more rapidly. The captain now concluded that the ship was lost and ordered her abandoned. With Pinto and two destroyers alongside, the troops were evacuated expeditiously and without resorting to lifeboats and rafts. Anthony's crew followed closely behind the soldiers. By 0905 hrs, the main deck was awash at the stern and she was listing badly. The last member of the salvage crew hit the water at about 1000 hrs, with Commander Gray soon following. At 1010 hrs Susan B. Anthony sank. No one was killed, and few of the 45 wounded were seriously hurt. All 2,689 people aboard were saved."

(USS Susan B. Anthony (AP-72) Wikipedia)

Northern France
Co. F 359th Inf., A.P.O.-90
26 June 1944- **Day 817**

Dear Zell and All,

Here I am in a foxhole in France. I can't tell you there what day I landed but I can tell you that there weren't very many who landed ahead of us.

We have spent nearly all our time on the front lines. I lost my gunner the first day in combat. A sniper got him. I think he will live O.K.

He made me think of Lehman Allen. He is a really swell boy, but that is the way with war. I am O.K. and feel swell.

Write often.

Love to all,

Leo

Note to reader: The 90th faced fierce, relentless battle. Leo reflected on this combat and how difficult this first month really was. He wrote this letter one week before he was wounded during a shelling and caught a piece of a German 88 tree burst. He was not alone. The 90th Infantry Division took over 25% of all Allied casualties in all theaters of war. The 90th sustained more than five thousand casualties in one week.

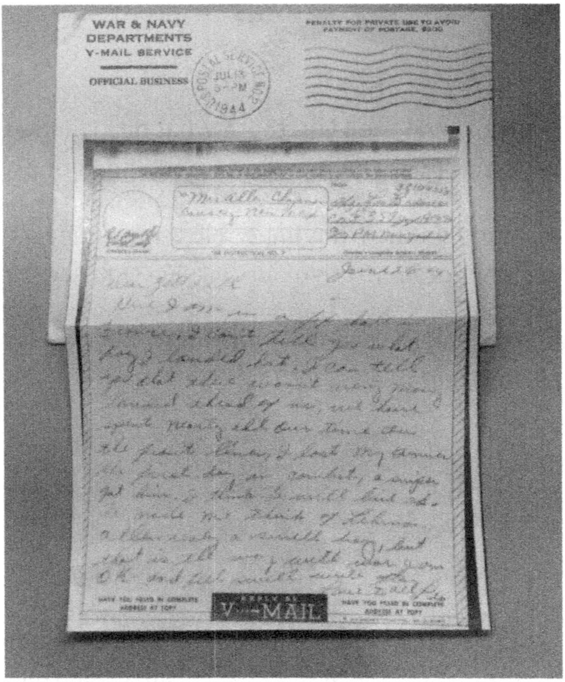

Letter From France

We didn't have to move to make contact. The Germans were in the same apple orchard as we were! They were behind one hedgerow, and we were behind the next one. In some places, we were two hundred yards apart and in one location as close as fifty. We didn't have any machine guns, but we had two 60mm mortars and G. Company had one of those. We were much too close to use mortars. We had no sooner started firing when someone from G. Company said, "We need mortar ammunition!"

Quinn Buffington, my mortar gunner, grabbed a load in each hand and headed in that direction. He hadn't gotten fifty feet until he was shot in the stomach.

I knew and was with that guy every day and night for nearly two years. I helped him get to the aid station. I thought he was dead when we got him there. The medics hooked up a bag of blood plasma. Pretty quickly color started coming back to his face. To make a long story short, he didn't die from this wound, but he stayed in a government hospital for nearly two years and never fully recovered from it.

The Germans had a lot of automatic weapons. We didn't know this until later in the day. They had hollowed all the back side of these high hedgerows and had little holes close to the ground to fire through. We would have to look over the bank of dirt, and they could look under most of it and be covered all the time. They were firing at us!

By night we had lost several men including an officer. We stayed up on watch: one on, one off all night. Word came down that we were to launch an attack the next morning at 7:30 A.M. There was a little Italian boy (American soldier) with me. We had picked him up as we came through England. Why, I don't know. Anyway, every few minutes he would ask me what time it was before it was 7:30. It got to be he would ask this every few seconds. By 7:30, he was rolled up in a ball in the bottom of the hole making animal noises.

I couldn't get him to come out. I was elected to take him to the aid station. When I got there, I told the doctor that he had cracked up and he pointed to a place for him to sit. There were five or six men there. They had eight hooked up to plasma bottles hanging from trees. There was a row of dead American soldiers; about twenty in one row. They were laying on their backs in perfect order. They had their O.D.

dress uniforms on. A lot of them were men that I knew for two years. Then, there was this smaller bunch, not so well lined up, and having their fatigue clothes on. No one was talking or had told me anything, but that is a sight that I will remember as long as I live.

I don't mind telling you that it was tough to go back to where the battle was raging. I don't claim to be a brave man because I am not, but I knew what I had to do and went back to join my company. The battle raged all day. We must have been taking a greater toll after we figured out that they had the hedges hollowed out and were firing from under them. The dirt was thin enough for a rifle bullet to penetrate it.

Lieutenant Raper received word that he was to assault the enemy and take the objective. There was a rifle platoon that had attacked it.

They only had one officer leading it, who was later killed. Lieutenant Raper and most of the men charged out to help them. Lieutenant Raper was killed. About 5:00 p.m., the Germans withdrew. Of course, we couldn't stand up until dark because we couldn't be sure. Pretty soon, some soldiers from the 9th Infantry Division started moving in among us. A lieutenant from that group told me they were going to make an attack and head westward and that we were to go southward toward cutting off the Cherbourg Peninsula. I wasn't sure we were going anywhere. I asked him, "Do you think you'll make it?"

He said, 'Yes, I think so."

I asked, "Why do you think so?"

He replied, "We fought them across North Africa and Sicily, and we haven't failed yet."

No one could ever know how much that raised my morale.

Frank Savarda, the only other New Mexican besides me in this unit (at that time) was nearby. It was getting dark, and we hadn't had anything to eat since morning. He opened his pack and got his little tin can of rations out, and each one had two bullet holes in them. He got a little nervous because he hadn't realized that there were any shots that close. After dark, a couple of boys went out in "no man's land" and got Lieutenant Raper's body. That said a lot about what his men thought of him.

Now, we had two officers left and we had to give one of them to G. Company to be their company commander (a job he kept until the end of the war). The other officer was our commanding officer (he was killed in just a few days). We had forty-two enlisted men left out of a company. I don't know how many we were short because we never got fully organized after the shipwreck and confusion at the landing. I always thought that if we could have all been together and missed those first two days, things for F. Company would have been a lot better; who was to know? This is the first time that I have ever told anyone about those first two days.

The next morning, the 9th Division moved through firing and never stopped. They turned West and we went South. We took a lot of replacements just for two days and it didn't seem real. There wasn't any way untried green troops like us could do what seasoned, battle trained, and tried troops could do. Don't ask me why because I don't know.

I asked this officer from the 9th Infantry Division quite a lot of questions. He told me we were going about it all wrong. He said, "Put pressure on the entire line and leave it light where they are fighting alone.

Then beat through the weakest part of the line. The Germans don't fight well when they are disorganized."

I never knew why we lost all those good men trying to take an apple orchard. We had one lieutenant left after we had sent one to G. Company. He was the one we were attached to while we were on desert maneuvers, when we went on night problems and had to find food and water buried in the desert using only a compass. He always came through. He was a Dallas attorney (and I bet a good one). I figured he would be the guy who knew whom to contact and inform why we got reduced to forty-two men in two days. There were a lot more. I know for sure that the division commander got relieved and his rank was reduced to colonel. He should have been relieved and shot! A lot of the old World War I retreads were cleared out and replaced with younger West Point men.

Things went better after that. There seemed to be an effort to tell us what we were trying to do and who was on our right and our left. Our mission seemed to be to continue going South until we cut off the Cherbourg Peninsula. It seemed like every day our objective was to advance a thousand yards. The artillery pounded us while we were on the move. We always reached our objective, but the price was always high. I guess there were other days just as bad as the first two days, but the casualties were not men you had been close to for nearly two years. We had such a rapid turnover that we didn't get to know each other very well.

We got two new second lieutenants. They were young and fairly small. One of them was redheaded; he looked like he might "fit the bill," the other one; however, had a peaches-and-cream complexion. He didn't look like he had done a day's work in his life. We were wrong about him. He was a doer and a stayer.

After a soldier had been there a week and met another who had been there two days, the one who had been there for a week was considered a veteran, and the one who had been there for two days a rookie. We were stopped one afternoon to let a new division pass through (more troops were arriving all the time). This young soldier came even with me, stopped, and started talking. He had been in Europe for only one day. I don't have the slightest idea why I made the remark, "If I were at the Cozy Corner Cafe, I would have a chicken fried steak."

He said, "You wouldn't be from Portales, New Mexico, would you?"

I told him, "Yes, I am."

He said, "So am I!"

That built him up. His name was Cleo Bryant. I think he was 18 years old and an infantry replacement. I think I saw him once after that, and then we went our separate ways. He wrote home that he had met me. I think his folks thought we were buddies and together all the time. The next I heard about him was when his sister wrote to me asking me how he had been killed. I didn't know he had until she had written about it. I always felt so sorry for those young infantry replacements. So many had died before they got any experience.

We kept moving south. We were receiving and delivering a lot of artillery. The casualty rate was high. Not as bad as other times but steady every day. We got help from the tanks during these days. I supposed that during training, we had the best tanks around. Our Sherman tanks had 75-mm cannons. They were not the high-velocity design like the Germans had. Somebody somewhere should have had to suffer for that! Our guns were like peashooters compared to the Germans. Since we had fought the Germans in North Africa with tanks, they knew our tanks were no match for theirs.

Tanks with this design were being sent over by the thousands. Again, they were no match for the 88-mm high-velocity cannons the Germans had. These guns could shoot through the front of our tank and exit the back. Our tank losses were four to one. That would apply to the loss of men as well.

A lot of guns could have been replaced between North Africa and Normandy. We captured a German anti-tank gun. The ammunition was laid out to fire. These shells, bullets, and all were about forty inches long. The butt that holds the powder was much bigger than the bullet.

Our tank shells were about 18 inches long and the same size all the way. When I saw this, I knew we had trouble.

We kept moving South until the Cherbourg Peninsula was cut off. We were then moved over, facing the East; there was a fairly tall hill to our left front that we heard a lot about. It was tall enough that they could observe what we did all the time. The only name that I ever heard it called was Hill 122. It was a must that it had to be taken before we could break out of the Normandy Hedgerows and get to the plains of France.

The hedgerows were originally built to hold back the seawater. Usually, they were rectangular in shape and spaced according to the land's level. The bank area between the hedges could be anywhere from a third of an acre up to ten acres. The closer to the water, the smaller the land area. They were built of dirt with a starting base about eight feet wide, sloping up with about a four-foot width at the top and four feet high. On top was a heavy growth of hedge, sometimes as high as ten feet, making a perfect defense for warfare.

Many times the Germans would hollow out the back side of a hedge to fire through it instead of over it. This furnished them with excellent protection. We replaced the 82nd Airborne Division. We were facing east now, getting ready for the big push to get from the observation of Hill 122.

We were there a few days waiting and getting ready until Cherbourg was taken. That would give us a good port and unloading facilities. A twelve-inch pipe was laid across the English Channel. We were in a defensive position. We kept an outpost out about half of a mile in front of the main line. There was a human-made drainage ditch, trees, and low brush under the trees. All movement was done at night. We had machine guns set up every two hundred yards. Under the brush that we cleared out was a three-wire barbed wire fence just three feet in front of us.

Odell Sullivan and I were on the outpost for twenty-four hours. We were told we couldn't retreat without battalion permission. We had no radio or telephone line. So, we could figure out that we were out there to alert the main line. We were sure to do a lot of shooting to survive. About 2:00 a.m., all of a sudden, the barbwire fence began to screech. Half of a minute later, it would do it again. We were sure it was German soldiers crawling under the fence. I whispered to Sullivan, "We could get the barrel of a machine gun over my shoulder, and you can carry the back, and we will start shooting as soon as we get close enough."

So, we sneaked through the brush and got all set to fire as soon as we saw the first German. We got a little further, and there was a black and white cow with her head through the fence scratching her neck. Needless to say, we were very relieved!

FRONT LINE

Back on the main line, we had our three 60-mm mortars set up in different firing patterns. I had to get up three times each night and fire the mortars. Even if we didn't hit any of their positions, which we must have, we would at least keep them awake most of the night. We took quite a lot of artillery fire while we were in the defensive position.

We didn't have to go on any patrols because of firing the mortars and manning the machine guns. They sent us more replacements. One day, we got two new second lieutenants. They both went out on patrol the night they arrived. One of them was really "gung-ho" and imagined that he would capture a whole gun crew. They were to capture any enemy soldier possible. The password for the night was "wagon wheel." About thirty minutes after they went out, we heard some running, someone really picking them up and laying them down and yelling, "Wagon Wheel! Wagon wheel! Let me through!"

It was the brave second lieutenant! We were later happy to find out that he was not assigned to our unit.

One night, there was a bunch (about ten) of our group who went on patrol. One got wounded, and they took him back to where another soldier and I were and asked us if we would get him to the aid station. So, we got a stretcher and took him to the aid station. As we were returning, we saw a patrol coming back, and they were spread out in battle formation. I asked my buddy, "What's the password?"

He said, "I don't remember. Let's hide."

I didn't think that was a good idea. If they saw someone hide, they would cut loose on us. I stepped out in the middle of the road and said in a loud voice, "What's the password?"

One of the soldiers told us, and everything worked out fine.

One day, one of our men visited the first platoon and came back and said, "Brown, we have a new replacement, and he says he knows you."

His name was Homer Mitchell and he is from Portales, New Mexico. I barely knew him; he was 19 and I was 25. I did not know him anymore by the time he was 17 years old. When you are that young, here is a lot of difference in age.

Homer Mitchell soon became "Mitch" to everyone. I had just gotten mail and had the Portales paper, which I shared with him. Mitch was a good soldier and a happy-go-lucky type of guy. I asked him about things in Portales. I asked him if he knew about a ratty place by the highway from Portales to Clovis. It was called "Boots and Saddles." It was always a thorn in the sheriff's side, a bootlegger's joint. Mitch said that he knew the place and that it had burned down. I said, "I bet someone set it on fire."

He said, "I know they did. The sheriff gave me a gallon of kerosene, a handful of matches, and a five dollar bill and drove me by and let me out.

He said, 'If you get caught, I don't know you.' "

Mitch told me that he never did get caught.

"The stage was set for the battle of the Forêt de Mont Castre which began July 3 south of the Douve. It was like the slow forcing of a massive iron door with hinges rusted solid by the Beau Coudray marshlands. To the northwest, at the other end of the 90th's sector, the door was locked fast by the formidable forêt and Hill 122.

Hill 122 was the eyes of the enemy. This 400-foot rise was a bastion from which Caesar's legions 2000 years ago repelled an enemy horde in the Gallic Wars. From a bald crest on the north, the forêt stretched to the south in a trackless and jungle-like growth.

For three years the Germans had fortified the forêt and had learned every inch of its terrain during maneuvers. The hill dominated the Cherbourg peninsula and keyed the entire southward drive to break out of the hedgerow country."

(Tough 'Ombres!" is a small booklet covering the history of the 90th Infantry Division. This booklet is one of the series of G.I. Stories published by the Stars & Stripes in Paris in 1944-1945.)

(note to reader: The 90th Infantry alone suffered over 5000 casualties during this week. That was 26% of all the casualties in all theaters of war.)

The word came down that the big attack was going to begin on July 3rd. Nothing was done to alert the enemy, but it was the 6th

Infantry Division in reserve in case of a breakthrough. The slugging it out and the breakthrough was up to the infantry. If they made the break in the enemy lines, the armor was to charge then through the hole in the enemy line and break them in pieces, disrupt communications and supply lines. If we got behind the German lines with fast moving tanks and armored infantry, mostly in half trucks, experience showed they gave it up soon enough. We were a long way from breaking through and taking Hill 122. Until we did, they could see everything we did in daylight.

The big attack was still on for July 3, 1944. Just before daybreak, the big guns cut loose. Then the medium guns started firing. The heavy water-cooled 50 caliber machine guns were firing over our heads and came down, hopefully, to get in their trenches. Things went pretty well until noon when the resistance began to build. By 2:00 P.M. we met heavy resistance from artillery and machine guns. They were well dug in and had people in trees firing down on the casualties, and we began to build up.

I was lying out in a level wheat field. The wheat was about two feet tall, and a machine gun was traversing back and forth over me. Every time the machine gunner made the sweep, the bullets would cut some of the wheat stalks off, and they would fall behind my ears and down my collar, but I didn't dare move. The only person I could see was one to my right front about 25 feet. I didn't know if he was living or dead, but I couldn't find out. There were lots of shells falling out in front of us, but none where we were. They had us pinned down with machine gun fire. They could have finished us off with artillery.

Two platoons of our company pulled a flanking attack, caught the Germans by surprise, and hit them on their left instead of in the front. We were laying out there as a decoy, and it worked. We lost a few men, but not very many; however, we captured quite a few German prisoners. They were dug in the side of a big ridge of dirt that had trees about 40 feet tall.

Several of them were up in the trees with machine pistols. They threw their guns down and began to yell, "Comrade, Comrade!"

We also found some Russian and Polish soldiers whom the Germans put on the front line and put the SS troops behind them, so that they couldn't retreat. They waved white flags and yelled, "Me Rusky" or "Me Polish!" I am sure that a lot of them weren't, but if they were prisoners, so much the better.

There was a German tank that kept shooting into the big ridge of dirt that we had driven the Germans out of. I don't know why it was shooting; the ridge was 15 feet thick. Anyway, we got two new second lieutenants. I liked the looks of one of them. They were just standing around as they hadn't been assigned. One of them just walked out beyond the ridge of dirt, and the tank shot him with the big gun. After that, we only had one new second lieutenant. His name was Lesser. *(Not long after that, on the 26th of July, the company commander was killed. Lieutenant Lesser took over the company and was made captain and was our commander for the rest of the war).* He was a good officer.

We dug in for the night. Then, I found out my good friend Odell Sullivan had been shot through the wrist. He was never fit for infantry duty anymore. I saw him guarding a bridge back behind the lines one time when we were in a truck convoy. I never got to see him again. He was killed in a car accident shortly after he got home.

The enemy withdrew about 200 yards and put up a new defense that night. We made a new attack early on the morning of July 4th. I was called to bring a mortar and knock out a machine gun position. But it was much too close. The mortar's minimum range was 400 yards. This target was only about 100 yards away. If I backed up 300 yards, I would be in a heavily wooded area. So, the commander said, "Let it go. We'll get the tank to do the job!"

When I got to where the rest of the platoon was, I learned that without any orders, they had fired all of their ammunition. All we had was what I was carrying. I just couldn't believe it. They could have been firing on friendly troops or whatever! So I called in, and the commander told me where we could get more ammunition. He said, "The firing you did was good! It hit right in the middle of a counter attack."

I tried to explain that I had nothing to do with it, but he told me, "Get five or six guys and bring up some more ammo of all kinds."

Things were getting hot! We went back about 100 yards. There was a depression about twelve feet deep with a straight up side toward the enemy and sloping in the opposite direction. There was one of our tanks behind the bank, along with a jeep that had just unloaded ammunition. The tanks had orders to move out and were moving around. That always draws artillery fire. We were just putting the ammo in our bags when we heard a deafening explosion just overhead of the six of us.

When it was over, one in our group was dead, two were seriously wounded, one was okay, and the other one and myself were wounded, but not seriously. It didn't hurt much. The shell fragment was very hot, and that may be why it didn't bleed more than it did. I took a shell fragment about the size of my thumb just below my right knee on the inside.

I took my gun sight off my belt, along with my compass, and we all walked back (there were a few more) about half of a mile to the battalion aid station. There was a doctor (a captain), three medics, and six stretcher carriers. The doctor told one of the medics to give all of us a morphine shot. I told him, "If it's all the same to you, I am not hurting that much and I don't think I need morphine."

He replied, "I have a policy: I give everyone a shot because a lot of people are in shock and don't know they are hurting."

In a little while, a jeep came up. It was called a "litter jeep" and had angle iron welded on the sides and across the hood, where they could tie down "litter patients," and guys like me sat in the back.

We went about two miles and came to a schoolhouse. It had seats all around the walls and a two-foot walkway on the outside with a rail about three feet tall. I was sitting on a seat inside, and all of a sudden, I got sick and crawled out to the walkway. I hung my head over the edge and began to "vomit up my socks." When I finished, a couple of guys rolled me on a stretcher and buckled me down. That must have been a good shot of morphine I got. I don't remember much of what had happened the rest of the day.

I remember riding in an ambulance. After a while, the ambulance stopped, and we were transported to another. Then, we arrived at a tent hospital, an "evac" hospital. We got a good meal, and I don't remember anything until the next morning.

The double tent held about 60 patients on folding cots. Each tent had a nurse and one ward. The nurse was a pretty rough looking old girl who did things the hard way. We were to be cleaned up before surgery. The nurse and the ward each took a pair of scissors and cut my clothes off. They took some Life Boy soap and gave me a bath. Then I got a clean pair of PJs. The nurse said, "We have got to get you a clean shave. The doctor will have a fit if he sees you looking like this!"

I said, "If I had something to shave with, I could do it myself."

She said, "That's my job. I will take care of it. All of this shaving cream and after shave is a bunch of nonsense!"

She took my Life Boy soap and lathered up my whiskers. They were long and dirty. She made no effort to get them clean first. She put her knee on my chest and her hand on my forehead and scraped my whiskers off. I hadn't ever had a shave like that!

The tent ward was full, and the surgery started with the ones who were hurt first. I was the last one. The operating room was a tent with four footlockers stacked two high on each, and two new 1X12 boards with a clean white sheet over them. Two doctors operated twelve hours per day and two more would then take over. These two had been at it all day when I went in.

There was a guy lying on the operating table; he was shaved all over. They were waiting for him to be moved before they started on me. I was sitting on a footlocker. One of the doctors said, "My feet are killing me. I have got to rest a minute and have a smoke!"

He took one drag on his cigarette and said, "Who am I to tell you that I have a rough life?"

The other doctor told him, "We took over a hundred pieces of shrapnel out of him, didn't we?"

The other doctor replied, "More than that."

I asked, "Will he live?"

He said, "Yeah, you can't kill these or kids; they are healthy, and we have the plasma to pour into them. We have this new drug, Penicillin, to keep down the infection. We don't lose too many at all that are alive when we get them."

He finished smoking his cigarette, put me on the table, probed around a while, and decided that they would cut up more muscles and do more damage than good by operating (I think they were too tired to do it).

They carried me back to the tent. The night nurse was on duty along with the ward. She was really pretty and had a clean white uniform, along with a nurse's cap. Instead of bossing the guys around like the other nurse, she would nicely ask, "Do you want to do this?"

I tell you, the other nurse could have taken lessons from her and saved herself a lot of work. At about 9:00 p.m. she woke me up and said, "Do you want to take a sleeping pill for me?"

I said, "Well okay, but I was already asleep."

She said, "There will be a lot of activity here tonight and you don't need to know anything about it."

I didn't. The tent had been completely refilled, and I slept through it. Later that day, they took me to a convalescent hospital. I stayed there about 36 hours, and the yellow pus was coming out of my leg like a rope. I showed it to the doctor. He had one of the wards bring me a pan of hot water and a towel. I kept a hot towel on it until it got cold. By the next day, it wasn't any better so he said, "I am sending you to England. We have far more than we can take care of here."

There was a landing strip on the beach about a mile from there. Close to it were huge tents set up and tied together. This was where patients were processed and left waiting for a C47 cargo plane. So, they started about fifty of us cross-country to this location. There were several planes coming and loading and several taking off. We got about halfway there and a couple of litter bearers yelling at the top of their voices. No one knew what they were screaming about, but they ran up beside me and said, "You are not supposed to walk under any condition!"

So they put me on a stretcher and took my shoes so I wouldn't walk. There must have been five hundred people waiting on the C47 cargo plane. They were carrying them out pretty fast.

There was an envelope with a string on one end of it that was tied to the front of our clothes. It was our medical records; up to date, along with the name and serial number. People were being picked up around me, but they never looked at my papers. After a couple of hours, one of the men said, "The rest of these S.O.B's are Germans and are going by LCT (landing craft tanks).

I raised up all I could and said, "Here is one who isn't!"

So, the guy came over, looked at my papers, and called to the other one that they had missed one. I was glad I wasn't stuck with a boatload of Germans.

We landed in London and an ambulance took us to a building, which I supposed was a hospital. The two men who took me in were dressed in white. Everything inside was white and spotless. They put me in a hospital bed and no one said a word to me. I don't know if they were American or English. After a while, someone brought me a tray of food and hung a urinal on the bed. After quite a long time, it got dark. I am sure I had been sleeping a lot because it was a very long night.

After daylight had come, someone had brought me a tray of food but still not a word. About 10:00 a.m., two guys got me on a stretcher and took me to an ambulance and onto a hospital train. We went to Birmingham to the 4203rd Station Hospital.

They put me in a hospital bed in a ward, and a nurse came over and took my medical package and tied it to the foot of my bed. She gave me a shot of the miracle drug penicillin and said, "You are not to get up for any reason

Two days later, she said, "Either you have disobeyed my orders, or you need a laxative!"

I had slipped to the potty a couple of times.

With the penicillin, the wound cleared up pretty quickly, but I was not to walk. The tendon inside my knee was partly severed. Every day, they X-rayed my leg and ask me about it, like "Had anyone removed anything from it, scraped it, or tried to remove anything from it?"

I told them, "No."

These questions came after a lot of X-rays. The X-rays showed the shrapnel two or more inches below the wound, along with the cut tendon. After about twenty X-rays, I asked the doctor about my wound. The shrapnel was below the injury; it came down instead of horizontally. I told him an 88mm "tree burst" hit me and the shrapnel came straight down. They couldn't believe how it missed my head, my entire body, and only hit my leg under my knee.

That, I couldn't explain either, but there was a band of dirt ten or twelve feet high between us and the direction that the shell came from. The fellow who was killed and the one seriously wounded both got hit on the top of the head. Anyway, that stopped the X-rays.

While recuperating in the hospital, I was lying in bed reading the paper and learned that the 90th Division moved to the 3rd Army from the 1st Army. At this particular time, the 3rd Army was running wild on the plains of France. The mission of the 90th was to seize the village of Ommeel and to take the high ground northeast of Chambois. Since the main road by which the German 7th Army sought escape ran directly through Chambois, the control of that town was vital to the allies, as well as the enemy.

The allied forces who severely trounced the German 7th Army also took a severe trouncing; the most complete and humiliating defeat ever suffered by the German Armed Forces. By this time, the men of the T-O Division knew they were "hot" and so did the Germans who were soon to face them again!

This hospital was made of metal Quonsets with big tents added on the back. The wound had cleared up and they moved me out back in a tent with folding cots. I was having trouble walking; I had to walk on the ball of my foot, and my knee was still quite oversized. They encouraged my walking.

I got a uniform and was allowed a pass to town anytime and with anyone. We went to town on several occasions. The guards were unarmed medics. I heard a couple of the guards talking one day. One said, "I'm not going to try to stop one of those infantry soldiers or the paratroopers. They might kill me!"

I started this story about the training and life of an infantry soldier. So, we will have to leave the 90th Division and follow the life of the soldier. History will tell of the war and the 90th Division.

My father got a card stating that I was relieved from active duty on July 8, 1944, due to wounds. They later sent my folks a card saying that I had been returned to active duty on September 27, 1944.

During my recovery, the 90th Division, along with several other divisions, broke out of the hedgerows onto the plains of France. July 26th was the worse day the 90th had. Our Company was about one hundred eighty-six men, and had over one hundred casualties; thirty-six of which were fatal.

After leaving Normandy, the 90th went North and was called on to help the Canadians who were close to the Falaise Gap. My friend, Stanley Skay, referred to it as "Hell Hole." They had the German army bottled up and took thirteen thousand prisoners, killing and wounding over eight thousand. I remember hearing them talk about the Metz Woods near Metz; the big stronghold. The 90th lost a lot of men there, and later, crossed the Moselle River during the flood stage (that is where I joined them again).

Two weeks after my arrival and recovery at the Army hospital they weighed me, and I weighed 130 lbs. When we left England for the invasion, I weighed 175 pounds (the most I ever weighed).

We got a lot of lectures from religious people and the military, mostly to pass the time. One person had been a missionary in India for over twenty years. He was interesting. He had mostly a hard life. One person who spoke to us was an Air Force pilot who flew a C47 cargo plane. He flew us back from France. He also flew supplies and sometimes paratroopers to the DZ (Drop Zone).

There was a paratrooper sitting beside me. He said, "They scattered us so far, and we didn't have a map to know where the others landed."

This trooper came unglued at the seams and was going to go up and kill this Air Force captain. Well, at the very best he would have got into a lot of trouble. About four of us piled on him, and the lecture ended, and the Air Force captain disappeared.

While in England recovering, I got to take some tours of the countryside to see things that had made history over the years.

RECOVERY

Since I started writing this paper, I have gained a lot of love and respect for this old country of ours. In the worst days, I never once had any doubts about what we were doing or if it was the right thing to do. I remember telling Odell Sullivan the day before the big push, "Odell, I wouldn't do what we have got to do tomorrow for a million dollars, but I wouldn't give anyone a dime to do it for me."

He agreed.

The U.S.O. had a large building leased in town. They and the Red Cross ran it. You could always get coffee, cake, or doughnuts there. There were tables and usually papers and envelopes so we could write letters. One or two times each week, there was a U.S.O. or Red Cross sponsored dance. There were always lots of English girls who came to it. It was well sponsored, and there was no booze.

However, booze wasn't a problem among their people or ours either. In fact, there may not have been any problems. At the pubs, you could always get bitters (and it was). I think it would have been beer with a little more time.

I saw people by the thousands, but for the first time in my life, I didn't know anyone that I saw on a day-to-day basis. I met this English girl at a Red Cross dance. At the end of the dance (11:00 p.m.), I asked her for a date. She told me that she was married to an RAF fighter pilot (Royal Air Force) and that he was shot down and killed over Germany. She said, "I get lonely, and I would like to go out with you, but if you want to go out with me for the reasons that I hear you Yanks go out with English girls, then I am not the right one for you, and I shan't change my mind."

I told her I would go out with her on that basis. The last thing I needed was to have an affair with an English woman who wanted to marry a Yank and go to America. There were a lot of them that did. We would go to an American movie once and a while, but I volunteered to take her to an English film. She said, "I just don't care about the British films."

We went to a pub, got a pint of bitter, listened to piped-in music, and played a few games of darts. And so the summer of 1944 went.

I was walking pretty well, but my knee was still swollen. I was called up for an exam to see if I was O.K. to be sent back. In the meantime, we had new people taking over the hospital, new doctors and nurses. They felt a lot more important than the last ones, who had been there for a while. The new doctor looked me over and said he thought that I was ready to be sent back. I asked him why my right knee was so much bigger than my left one. He said, "I don't see any difference."

He went and got a tape measure, measured them, and didn't say anything.

I asked him, "How did they check out?"

He said, "One is three inches larger than the other."

I was lying on my back, and my right knee wouldn't straighten all the way out. He felt them. Directly, he put the heels of both of his hands on my right knee and quickly put all of his weight on it. I yelled 'till you could have heard me two miles away! I drew my other foot back and nearly kicked him in the face. He turned around and left and never dismissed me or said a word, so I stayed there a while, and then left. He had on glasses and, if I had kicked him in the face like I almost did, I would probably still be in the stockade.

This new bunch who took over the hospital was a lot more for rules and regulations than the older ones. They decided that we should all pull a parade. The doctors were first with a major in command, next were the nurses with a male doctor in charge, and then came the ward boys. The rehabilitated casualties were in front, but got off the roadway to bring up the rear. We were made up of mostly infantry and paratroopers and, if there was one thing that we knew how to do, it was to march in a parade.

The doctors, nurses, and ward boys came by about half out of step. This bunch of casualties was led by a first lieutenant of the infantry. When they came by, they were about a third out of step; bobbing up and down.We got tickled and laughed out loud. The major stopped everyone, stomped back to where we were, and asked the lieutenant in charge of us," What's so funny, and why are you laughing?"

The lieutenant said, "Sir, I don't know what they are laughing about unless it is the way you are marching. I have no idea what is wrong with that."

He started busting out laughing. So, the parade went on. I figured it was the last one.

One afternoon, just before chow-time, we were told we would be going to a replacement pool. The next morning, we were called out and checked off. Several were not available, so they had a mark drawn through their names. We boarded a troop train to this huge replacement pool. It was called, "Colonel Kelvie's Concentration Camp." He was the ex-general who was relieved of duty by our division commander and reduced to the rank of a colonel.

We drew an M-1 rifle, cleaned the grease out of it, and test fired it; mine was a "beaut"—a good one. We also got clothes, packs, and the works. We had both a full field pack and duffel bag. I drew guard duty (when you move to new place, they always had a duty roster and started at the top. If your name started with A or B, you would always be called up). I figured that if I knew how to do anything, it was guard duty. We had a Sergeant of the Guard who could swear about as good as our first sergeant. He thought that he was tough. He cursed and ranted and raved for some time. I wasn't impressed at all. He said, "Before twenty-four hours are over, I will have a bunch of you in the guardhouse. I will slip up and grab your rifle off your shoulder and turn you in for letting someone have your rifle while you were on guard!"

My thoughts must have shown on my face. I neither said anything nor looked up at him. All of a sudden, he stepped up right in my face (he was about my size) and said, "You think you're pretty damn tough?"

I said (really low so only could hear it), "No, I am not very tough, but if you come to my guard post after my rifle, you are going to get in more trouble than you can handle."

He stepped back and didn't say anything because I think he knew I meant every word of it! I never saw him during the twenty-four hours of my duty and am glad I didn't."

England (4203 U.S. c/o Postmaster, Army Hospital Plant)
19 July 1944- **Day 834**

Dear Zell and All,

How does this find everyone? I am O.K.

I am back in England now in a G.I. hospital. I got a small piece of German 88 shell in my right leg. The place where it went in has healed up. They say it won't bother me to leave it in, but I sure want it taken out!

I can walk O.K. but I have to walk on the ball of my right foot. It seems that muscles are too short to set my heel down. It doesn't hurt at all now. It was just about O.K. before I came back to England. It sure does seem quiet over here to compare with France.

There is a Px where we can get our rations (3 bars of candy, 3 packages of cookies, and a pack of gum + 2 Cokes) per week. I never have money to get a Coke, yet, I have had one pop since I have been gone from the States—it was a Dr. Pepper, just before I went to France, in an American Red Cross.

I spent 28 days on the front line and that is a long time. The good Lord was with me and I was....

CENSORED

There were three of us there and I was in the middle one and I got off the lightest and that is too much to just be called luck.

I came back to England in the transport plane. It was a real nice ride and just a matter of minutes by air.

I rode on one American ship, a British ship, and a landing barge going over there.

They really take good care of the wounded. The doctors and nurses are O.K. and my hat is off to those first aid boys on the front lines. They are on the ball.

Write and tell me how things are going at home. I haven't had any mail in a long time.

Love to all,

Leo

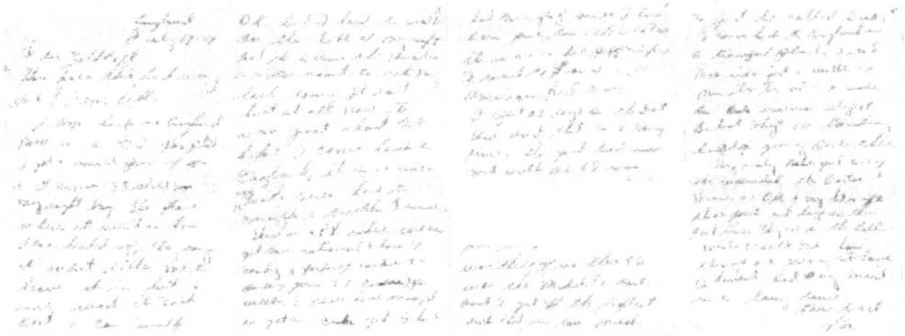

4203 Army Hospital_ 19 July 1945 Just Wounded

4203 Army Hospital Plant, England
Co. F 359th Inf., A.P.O.-90
1 August 1944- **Day 853**

Dear Zell and All,

How does this find everyone? I am O.K.

The news sounds good over in France. I don't see how the Germans can keep going the way that are losing men and material.

Boy I hope that this thing is over soon.

How are things going around Causey these days? I haven't had any mail in a month now. Some of the boys got some today. Maybe I will tomorrow.

I noticed in the paper today that the 90th Infantry Division is still on the line: They ought to give those boys a break some day--it looks like.

I saw a good movie today, "What Happened in Mable's Room" It is a comedy and is really funny.

The British put on a stage show last night. It was pretty good for a limy (English) show. Some are different from the ones you see in the States. The girls wear more clothes in a stage show.

It helps to know they want to cheer us up.

Anyway, write soon.

Love to all,

Leo

4203 U.S. Army Hospital Plant, England
Co. F 359th Inf., A.P.O.-90
15 August1944- **Day 867**

Dear Zell and All,

How does this find everyone? I am O.K.

I have got a lot of my back letters but no package yet. I got your letter, the first one you wrote to this address. It was the first one written directly here.

Things are O.K. by me. I saw a good show yesterday, "Hi, Sailor."

I like any kind now, except a war picture and I don't ever want to see a war picture again. I guess they realize that because about all we see are musical comedies.

The news sounded very good over there now.

I haven't had any newspapers from home since I left France.

I got a letter from Delbert yesterday. It was to this address.

How much longer did he thing he would be in Kentucky? His outfit never has been on maneuvers. Have they?

He must think they will be there quite a while judging from him taking the car back down there.

I am afraid he is spending money a lot faster than he is making it on that car, don't you suppose. I don't know much about it, but I know he never had that much money after a furlough. Ha Ha

Write soon with all the news.

Love to all,

Leo

U.S. Army Hospital Plant, England
Co. F 359th Inf., A.P.O.-90
30 August1944- **Day 882**

Dear Zell and All,

How does this find everyone? I am O.K.

I have moved again as my mail is slowly catching up with me. I had a letter from Glen today. He said they had just had a 3-inch rain. I hope you got as much. Will the crops be good this year? They have just got done cutting wheat here.

I have been to town a couple of times since I have been here--the first time since about two weeks before D-Day.

I went with a girl out of the R.A.F and we had been to the cinema. A couple of female British M.P.s stopped us and chewed her out for wearing silk stockings. They took her name and organization so they could report it.

Can you guess how you would feel in a place like that? About all I could do was stand and chew my gums.

After they released my dame we started on down the street and I said, "Are things Okay now?"

She said, "I guess so. I still have my Yank and that was why they griped me out; they didn't have one."

By the way, "Gone with the Wind" has finally come over here.

I saw another good show, "Hold back the Dawn."

There is quite a mixture of soldiers in town. There are more Limeys (British) than any. There are also soldiers from America, Canada, Australia, New Zealand, Free French, Poland, and you also see Italians and Sicilians who are prisoners but the British really don't bother to guard them.

I think the Australians have the best looking uniform of them all. It is as dark as our Navy uniforms only they are cut in Army style.

I have about run down, so Bye for now.

Love to all.

A Bud,

Leo

U.S. Army Hospital Plant, England
Co. F 359th Inf., A.P.O.-90
14 Septmeber1944- **Day 897**

Dear Zell and All,

I just received your letter of 24 August. I got the box of chocolates you sent with the stationary. It was all in good shape and really tasted good.

You asked me what I needed. I really need a razor and shaving brush but I don't know if it would be worthwhile to send it until I get a more permanent address.

Mail service is poor for you are always moving from one ward to another, then from one place to another. I expect the replacement pool will be my next stop.

You can tell Don that people over here don't know what watermelons are. We just got done harvesting the hospital's potato crop of about 3 acres. A couple of English girls plowed them up and we picked them up. My legs are sore from it.

I just came from the show, "Lady in the Dark." It was an old show I had seen a long time ago.

I just heard the news. They said our 1st Army was 10 to 12 miles inside Germany and that the U.S. 9th Army was in France. That makes 4 Armies the U.S. has in France.

From what everyone says, the crops must be pretty good there this year. I hope so. I wish I knew I would be there next year. From the looks of the demobilization plan, I can't see anything to keep me from going to the Pacific. Those aren't pleasant thoughts but I had as well expect it, then if I don't go, that will be grapes.

I don't know what to think about when the war will be over here. I think a lot depends on how long it will take to break that west wall line. Then, if the Germans decide to fight to the last or give it up as a bad job.

I wonder if the German people can take it when they start shelling down every house the come to. I don't expect they will go to much trouble to save their property. I am sad there are several men lost that way in France. We tried to save all the French property we could.

I will hand it to the French people. We would come up on them where most of their stock had been killed and their house torn down, but they usually were happy and had a smile on their face.

I am sure that where the troops went so fast, they didn't tear up much.

There are quite a few Free French soldiers in this hospital.

Write often,

Love to All

Leo

U.S. Army Hospital Plant, England
Co. F 359th Inf., A.P.O.-90
16 September 1944- **Day 899**

Dear Allen,

I received your letter of 26 August a couple of days ago. I was glad to hear that you had a good rain and things looked fairly good.

Things are about the same over here. We had a little less rain this last week. Planes have been able to fly most of the time day and night. I expect Hitler has got a lot of bombs in the last ten days. The American planes go and come all day and the R.A.F. at night.

There is a new bunch from the states taking this hospital over. They have jut been in England a week. I talked to a couple of them last night. They said, "I don't see how you stand this weather over here?"

They said they had only been here a week.

I said, "The last ten days is the about the longest period of fair weather we have had." I said, "If I had a place like this to work where I knew I had a dry bed to sleep in every night, I wouldn't worry about the weather."

It looks like I will be going back over anytime now. I hope the 3rd Army has that line broken before then. I hope I can get back with my old Company.

I spoke with a guy that had just come from the Company a few days ago. He said my old Commander was back with the Company now. He got hurt the first day we were in France so a 1st Lieutenant took the Company (he had been with us a long time) about a week later, he was killed. So, we got a new Captain none of us had ever seen before. He wasn't a man for Company Commander.

A good Captain can really help you a lot and a sorry one can get you a lot of grief.

It looks like when the war will be over (over there) is anyone's guess. I think a lot depends on Russia and if the Germans give up or fight to the last--Don't you think?

I don't see anything to keep most of us guys over here from going to the Pacific.

I haven't been to town in over a week. I guess I will go tomorrow. The blackout will be off then. It will be a dim out.

Did you all ever go out to see Dale and Iva?

Answer when you can. I am always glad to get a line from you.

As ever,

Leo

U.S. Army Hospital Plant, England
Co. F 359th Inf., A.P.O.-90
24 September1944- **Day 907**

Dear Zell and All,

Howe does this find everyone? I am O.K.

I received your letter of 4 September. I was glad to get it and also the clippings out of a local paper. It gives you an idea that on the "Lend Lease" that we don't give all and get nothing in returns. Though, if it were dollars and cents it wouldn't stack up too much to compare with what we let them have.

From reading the papers you would get the idea they don't think as much of you as they did before they were sure that Germany was about whipped. You know, they won't need us much longer (I hope). The paper doesn't say everything in plain English but generally you can get the idea if you know what I mean.

I think they are jealous of America's industry. They know we have them beat by a city block. I read an article a couple of days ago. The title was, "What is the Mater with Britain's Industry?"

It said an automobile made the U.S. would cost, if weighed, 15 cents per pound and if it were made by Britain it would cost 29 to 30 cents per pound. It went on to ask, "Why would this be when the workers in the U.S. make over twice as much as they do in Britain?"

You said you read the English paper and couldn't tell anything about prices. I will try to explain that 6 pence is the same as a dime. 9d would be 9 pennies and same as 15 cents: If it were written this way 3/3 or 4/5 that would be 3 shillings + 3 pence, equal to 65 cents. The other would be 4 shillings + 5 pence is about 89 cents. A shilling is worth 20 cents and 12 pennies make a shilling. 3 pennies equals 5 cents

A price could be written either of these ways and mean the same 2/4 or 2S + 4d- that would be 2 shillings +4 pence.

You wouldn't have any trouble counting the money. The hardest part is understanding what they say when they ask you for money.

You asked me what I would like for you to send. I need a comb, hair oil, tonic, shampoo, and aftershave lotion. I can't buy any of these here. Make them all small bottles or they will be heavy and I would be able to carry much around because I probably won't be in this place long. By the time I get it, I might be back in the fight and not have any use for it.

But, from the way most packages get around, the war may be over by the time it got here.

I told you in my last letter I needed a razor and a shaving brush, but I don't need them at all now. They sent my shaving articles to me that were rolled up in my blanket in France.

Write soon,

Love to all,

Leo

U.S. Army Hospital Plant, England
Co. F 359th Inf., A.P.O.-90
6 October 1944- **Day 919**

Dear Zell and All,

How does this find everyone? I am O.K.

I just got off a 24-hour guard. It was not bad. It will probably be 24 hours per day soon anyway.

I was glad that you and Mom got to go to see Iva. What do you think of that Country? A shade rough no? I have been to Clifton and Morenci, Arizona before and way up north of there is a lot of beautiful country, tall mountains, and beautiful forests.

I never did get the gloves you ask about. I got a box of candy with stationery in it. I sure hope I get the gloves soon.

I had a letter from Glen today. He said things look good there.

I have seen the show, "In Our time." That old lady that eats all the chocolate was just like the English and that girl also, for that matter. Fall fast for a stranger—that is them.

Did you see, "Bachelor Brother?" It was a good, or maybe it was "Bachelor Sister."

Yes, I got the letter from Cordia. She told me about her baby being born dead. I don't guess Art got there until later, did he?

I had better close and write a letter to Delbert.

Love to all,

Leo

P.S. Tell Allen and the boys hello for me.

Army Ground Forces
Replacement Pool, England
Co. F 359th Inf., A.P.O.-874
8 October 1944- **Day 921**

Dear Zell and All,

The enclosed $12.00 money order is to buy Christmas presents for the members of the family. It isn't much but it is all I have, so maybe you can find something that will go around (I hope).

We are having another nice rainy, foggy day.

Nothing is different here.

News is short. You can tell them I thought of them anyway.

Love to all,

Leo

France
21 October 1944-Day 934

Dear Zell and All,

How is everyone back in the good old sunshine States. It is really raining here and muddy-Oh boy!

I just read in the paper that they had invaded the Philippians. Boy, I hope they whip those Japs soon. I really would hate to go over there from here.

I have seen quite a lot of France that I never saw before in this part. It isn't as torn up too badly. There are many a train shot up between here and the beech. The railroads are in pretty good shape now. We have a lot of trains running but they are slow. That was a tough ride from the beech to here (three days on a boxcar without room to straighten out our legs or lay down. I really was glad to get off. I have received a couple of letters since I have been here.

We have a long way to go to the frontline yet this would be nice country if it only wouldn't rain all the time.

Did you get the money order I sent just before I left England? Delbert wrote that Ray Altman was in the hospital. He didn't know if he was wounded or had malaria. He had it once and was in the hospital in Australia.

Write and tell me all the news.

Love to all,

Leo

France

November 20, 1944-Day 964

Dear Zell,

I have time for a few lines. I am just fine. How is everyone? Well, we are having some more of the fine rain, as usual. We had three or four nice days and that sure did help. We did okay in the pretty weather.

I guess everyone is really busy there this time of year.

How did the crops turn out this year?

I haven't had any mail in a long time. I had a letter from Jimmie Williamson. He wrote to the hospital-said he was coming to see me.

Love to All,

Leo

Germany
Co. F 359th Inf., A.P.O.-90
1 December 1944- **Day 975**

Dear Zell and All,

How does this find everyone? I am just fine.

I just received the letter that you wrote me while I was in the replacement pool in England.

I guess Allen is through combining by now. Did he pull it with the regular this year?

The way Glen wrote, he though they might buy a combine. I don't know if they bought one or not.

You said you bet I wish I was there working instead of here carrying a gun. You are just not kidding. The hardest days I have ever seen there would look like a picnic to me now.

I am sending a little clipping about the 90th out of Stars and Stripes paper. I would you would save it for me? It ought to answer the questions you ask me.

By the way, did you ever get the little money order that I sent just before I left England? We were supposed to get paid today but now I don't know.

I don't know if it will be in Marks or Francs. It doesn't make any difference. I can't spend it anyway.

I never did get the gloves the folks sent, but I have a pretty fair pair of G.I's now.

Write soon,

Love to all,

Leo

P.S. Tell Allen and the kids hello.

COMBAT

Two days later, we loaded on an English ship to cross the channel. We had just got out of the post when an American soldier came to our compartment and called off names (starting with A and B) for twenty-four-hour K.P. To make a long story short, I had enough of that system for duty rosters, so I didn't answer. This guy asked someone, "Do you know Leo Brown?" and he said, "No, unless..."

He pointed at me.

I told him, "My name is Leon Jones."

He said, "If you see this Leo Brown, tell him to go to the kitchen."

I said, "I don't know one person on this ship."

That was the truth. I went up on top of the deck until bedtime. When I woke up the next morning, this same fellow was yelling for Leo Brown. About that time, the plot thickened. It came over the P.A. system in a British accent that they had found a helmet, shell belt, and gas mask out by the deck rail. One of the crew saw a man jump overboard. Since no one could find Leo Brown, they figured he had jumped overboard.

We dropped anchor at Omaha Beach at a makeshift dock. Some first sergeant was calling names as we marched off (starting with A). Pretty soon they got to Leo Brown and didn't slow down because they knew Leo Brown had jumped overboard. But I answered. This first sergeant came to a stop and said, "Where in the hell have you been on this ship since yesterday?"

I said, "I have been on top of the deck most of the time."

He looked frustrated and said, "Who in the hell jumped overboard?"

I said, "I don't know, I didn't see anyone, and if I had, I wouldn't know who it was because I don't know anyone on the ship."

I still don't know if anyone jumped overboard or not.

I got a good look at Omaha Beach. It had a sea wall at least 100 feet straight up. There had been active gun emplacements up and down the front, facing the sea. Our boys from the 1st and 29th Infantry Division had to take these on D-Day. Why they landed there, I do not know. It was hard to climb without facing an armed enemy. Everyone from Eisenhower down thought that the bombs and Naval gunfire had destroyed all of the enemy guns on the landing zone, but they hadn't been destroyed. There were enough casualties in each of the armies to fill a train.

We assembled on a muddy piece of ground inland about half of a mile. We ate C-rations, and the next day we were split into armies. I was put in the third army now. We were loaded on trains into what was called "forty-and-eight cars."

In World War I, they got this name because they transported forty men or eight horses. In our car were forty enlisted men, two officers, six cans of water, and six boxes of rations. We all had a full duffel bag and field pack. I was lucky. I got close enough to the wall and could lean against it and hang my pack on it.

It was a long night, and a few fistfights broke out. Finally, morning came, and some of the men got on top. We could open the side doors, and we had a little more room. We got into Paris, and there were three open-top freight cars next to us with good Army rations. We finally convinced the officers that it would be O.K. to trade our C-rations for some good ones. That wouldn't be stealing. We got to another tent camp and were divided into corps and then regiments.

Each time we would stop, we would lose two or three guys. They would get too far from the train and couldn't make it back in time to board the train. They would later be arrested and put on the next train. Someone had to guard them, so that person had to carry two rifles, and the prisoner carried none.

While we were at the regiment replacement center, we tagged and turned in our duffel bags (and never saw them again). We went through a line and got a carton of cigarettes, a day's K-rations, and the colonel was passing ammunition and would say a few words to each soldier. One of the boys who was a prisoner and had no rifle came through and the colonel said, "Son, where is your rifle?"

He said, "Sir, I am a prisoner and don't have a rifle."

The colonel said, "Why, you look like and American."

The boy said, "I am, sir."

The colonel replied, "The only kind of prisoner we have are Germans! So, give this soldier a rifle!"

There were never any charges filed. After dark that night in heavy rain, there were two guys who came up and said, "We are here to take you to battalion headquarters."

It was so dark that we had to hold on the pack of the guy in front of us to keep from losing our way. It wasn't very far. When I got there, I finally saw someone I knew at regiment headquarters – old Stanley Skay.

He had been injured once in a jeep accident and wounded twice in battle. We got to battalion headquarters, and there was a big barn with hay in the loft. We were told to stay there so that in the morning we were to get breakfast and wait until someone from our company could come by and get us. It was nice, dry, and warm in the hay barn. The next morning I was first up and climbed down the ladder. There were eighty German prisoners of war sleeping there and only two Americans, a first lieutenant and a captain from G2 (intelligence). I said, "If you had told us, we would have furnished a guard."

The first lieutenant said, "They are harmless. I talked to every one of them. They are finished and glad to be out of the war."

He said, "I don't mind that. I can get just as much information out of a bunch if they lie as if they tell the truth. They always make a pattern. If they lie, they will spend so much time dodging the truth, and that is very obvious."

He just sounded like someone who knew what he was talking about. I saw him a couple of times more.

Skay and I were sitting around a fire drinking coffee and catching up on talk, and I looked over where I had leaned my new shiny M-1 rifle, and someone had traded me for a beat-up old rusty rifle, the worst I had ever seen.

“On the 8th of the month, the Regiment assembled in secrecy on the west bank of the Moselle, near Cattlenom. By daylight of the following day the First and Third Battalions had crossed the river in assault boats, and begun the attack on the enemy's river defense. The Second, crossing later in the morning, experienced difficulty in finding necessary boats and was subjected to Jerry's observed artillery fire. Meanwhile, the once serene and peaceful Moselle broke into a raging torrent whose flood waters threatened to overrun the entire landscape. Bridging operations by the Engineers came to a standstill. At the same time, the enemy plastered the entire river line with all the artillery he could muster. The situation fast became critical. Medical supplies, rations and ammunition had to be rushed across by boat. On the east bank of the river there were no vehicles, and all supplies were hand carried to the fighting men on the line.” (Peragimus, "We Accomplish"A Brief History of the 358th Infantry)

I was down by the Moselle River three or four days after the crossing. There was a jeep leaving and someone standing up on the passenger side. It was none other than General George Patton! We had been across the river with little food and no blankets except the one that we carried when we crossed the river when it was high and in flood stage. Now it was back in its banks, but still there was no bridge. I didn't see or hear this directly, but General Patton told the colonel (engineer) he wanted a tank across the river in thirty minutes. The colonel asked how he thought he could get it over there (I got this firsthand from the captain who was there). Patton said, "I don't care if you have to fly it over there."

The colonel replied, "That's impossible."

Patton said, "You are relieved!" So, Patton asked the second in command, "Can you put a tank across the river in thirty minutes?"

He said, "I can."

I couldn't believe my eyes when we walked up and there were two pontoon boats with a big steel bridge across them (The two pontoon boats had a huge steel bridge made of twenty-four inch channel iron laid across them, thus making it possible to drive the tank on them, and then, the jeep was just ferried across the river). There was a tank, a jeep, and an anti-tank gun behind it. There was a little outboard motorboat loaded with equipment for unloading. The tank made it across and was unloaded in twenty-five minutes!

Everyone except F. Company was getting a lot of

counterattacks. We found out when we went to move that there was a big minefield directly in front of us. Stanley Skay and I got a roll of blankets each and were going back to our platoon when we got another shock. There was a German tank coming down the road to meet us, and there were a bunch of American soldiers standing all over it. They were yelling, laughing, and carrying on. When they got to the river, a captain told them, "You had better park that until you get a white star put on it!"

I never heard anything else about it, how they captured it, nothing.

I took my roll of blankets to the mortar section, and Stanley Skay took his to the machine gunners. In less than thirty minutes here, old Stanley came, limping over where I was, and said, "Brown, I need you to help me to the first aid station." He had a piece of shrapnel in his jaw and in his leg (he went to the hospital for six weeks for that).

We began a campaign to get into Germany itself. There were many pillboxes. We found that if we came up on the blind side of the pillbox and walked on top of them at night, we could drop a white phosphorous grenade down the vent pipe and could get the occupants out without much trouble. We didn't have enough men to occupy them, so someone else would take up position again after we were gone.

A "pillbox" is a military fortification made of steel-reinforced concrete with walls six to ten feet thick and two or three gun ports. They contained an apparatus that could operate to get fresh air inside. This was the only straight-up opening to the pillboxes, which made it very handy to drop a white phosphorus grenade. The phosphorus would melt through the metal making a terrible smoke, which would flush out the enemy.

About this time, I got a three-day pass to Paris. The Red Cross had leased two hotels from the French and rented the rooms to American soldiers. The guests could also get food there. The total cost per day, I think, was ten dollars. We got clean clothes and a place to clean up as well as a haircut. One night, I was in bed in the hotel room, and I heard something rattle the door. I reached under my pillow and got my 45 automatic pistols and set up in bed. Just then, the lights came on, and a captain and staff sergeant stood in the door. Both with their hands raised. I said, "You're going to get somebody killed doing that!"

The captain said, "I think you are right! Will you please put the hammer down on that cannon? All we want to do is see your pass. There are a lot of men absent without leave in Paris."

There was no way you could rent a room without a pass, but I guess you could get a pass for a price.

The French men in Paris (it seemed like all of them) carried briefcases. They were buying cigarettes and selling French money for three-to-one for American or English money. They seemed busy. There was a guide for a fee that would take soldiers on a tour of Paris. We went to the Eiffel Tower, the Arch de Triumph, Napoleon's tomb, a huge church, and a lot of points of interest. The three days there were up too fast.

When I got back to the company, they had been in reserve or put on hold all the time I was gone but were ready to move to the Saar Valley and on toward German soil. We got word from the G2 people (intelligence) that there were two pillboxes ahead of us that had some high-ranking German officers they wanted to capture. They said there was a relatively large minefield. The minefield was marked, and we measured it. None of us could see anything to indicate a minefield. After we had got past it, we sat down and waited for it to get good and dark. The boy was it dark! The pillboxes that we were to take were about half of a mile from the supposed minefield.

It was so dark that we couldn't find the pillbox. We held hands and continued to walk. Then we found it and caught our guard (there were two we were to take) nearly asleep. He didn't give us any trouble. The guard at the other pillbox was more alert and ran inside and closed the steel door and called artillery on the pillbox. That killed one American and wounded one. When we dropped a grenade in the vent pipe, they were sure ready to open the door and come out. We captured a full colonel, a regiment commander, and a battalion commander. The intelligence people took over and got a lot of maps and a lot of goodies. They found good maps of the layout of the Siegfried Line. This was the first time our people knew to what extent it lay.

The Siegfried Line was a German chain of steel forts and concrete barriers that extended along the border between Germany and France. It was completed in 1938 under the supervision of Adolph Hitler and was designed to make it impossible for attacking forces to get through. The Allied Force assaulted and broke through the line in many places. Later, it was destroyed.

Hitler would probably be horrified to learn how this line was originally broken. A bunch of ragged-ended American GIs gathered up all the mattresses, quilts, and blankets they could find and walked across Hitler's impenetrable Siegfried Line! He never considered the "good old American ingenuity." Where there is a will, there is a way.

We moved into Germany. The artillery fire was really bad, but we had some real sturdy houses to get in. The next morning, the company jeep had several shrapnel holes in it, and all four tires were flat. Shortly, Service Company had some more tires and wheels on it.

We were to join the 95th Infantry Division, cross the Saar River, be joined by a couple of other armored divisions, and make a big drive, circle north, and hopefully trap a bunch of Germans. We pushed on toward the Saar River, and we had gotten a lot of rain as usual. The 357th and 358th Regiments crossed the river but had run into a lot of trouble. They bypassed several pillboxes that were occupied. Along with many counter attacks, the artillery fire was fierce! This added up to the fact that there wouldn't be any big new offensive attack. It was a matter of holding on to what we already had.

Siegfried Line

"The 2nd Battalion, crossing after daylight on a footbridge, faced heavy mortar and machine gun fire from enemy pillboxes commanding the river line.

Once across the battalions faced the fortified cities of Pachten and Dilligen, bulwarks of the Siegfried Line. The main thoroughfares took on names like "88" Street and "Purple Heart Avenue". It became a war against steel and concrete. Troops were raked by fire from pillboxes cleverly concealed in harmless looking barns and shops. Once again supply men and engineers battled against a river. The Regiment's supporting tanks and TDs and the guns of the Anti Tank Company, had to be ferried across, for Jerry's heavy shelling prevented the construction of a bridge. Supply men pushed supplies of ammunition and rations across on every type of boat; alligators, ducks, assault boats and storm boats. Wiremen struggled day and night to maintain communications across the river despite a swift current and constant shellfire. For sixteen days and nights the battalions hit again and again into the enemy's fortifications. Continuously hammered day after day, the enemy was systematically blown out of one pillbox after another, as all three Battalions were employed to clear the major portion of Dillingen." (Peragimus, "We Accomplish"A Brief History of the 358th Infantry)

At the end of the second day, word came down that our regiment, the 359th, would cross that night. Our battalion moved to about half of a mile from the river, and we ate supper. My feet were hurting me and were swollen. I went to the medic. The doctor saw me and painted my feet with something purple and said, "Boy, you are going to have trouble with those feet."

Captain Lesser told us, "After we've crossed the river, go on straight east on the right side of a railroad embankment."

We heard a big shell coming, and everyone hit the ground. I was in a shallow barrow ditch. One of my friends came flying over me, and I asked him," Are you hurt?" (I just knew he had to be.)

He said, "My shoulder went numb, but it doesn't hurt."

I felt him; there was a big piece of shrapnel in his pack and it was really hot, but he wasn't hurt.

We found out shortly that the battalion commander was killed along with seven or eight others and there were twice that many wounded. I sent one of our new replacements as a stretcher-bearer back to the hay barn where we ate supper. I told him to come right back. We crossed the river shortly. After about half a mile, we stopped and found we had lost contact with Captain Lesser. He had continued down the railroad embankment not knowing what had happened to the rear of the column.

There were two new second lieutenants with this later bunch of replacements. There was a trestle over a road turning north under the railroad. We were supposed to stay on the right side of it, but the lieutenants turned north under the elevated train. It was very dark, but the rain had turned to snow. This lightened things up a little bit. We went north until we came to a fence and turned around and came up on a pillbox. There was someone standing in the door of it. One of the second lieutenants said, "What's the password?" We got it.

A burst from the machine pistol (similar to our machine-gun only smaller and able to fire more rapidly). Someone, not an officer shot the German guard, and someone else started throwing hand grenades. There wasn't any way of knowing where they were coming from. So, we got out of there and went a couple of hundred yards and pulled out a radio antenna, which if they were in doubt, they should have done in the first place. They contacted Captain Lesser and asked," Can you hear the firing?”

He said, "Yes I can! Go back under the railroad and go on east like I told you to do to start with!" He wasn't real happy with them. We were lucky we didn't have a lot of casualties!

It had been a long and terrible night. We finally got to where Captain Lesser was, and he told us, "Go south along the trench. Everywhere there is one man, leave one of ours to help strengthen the line."

I was so tired and wet and miserable; I didn't think that I could make it. But when I got to this post that had one man, I knew I could. There was an 18-year-old boy who had just come over as an infantry replacement only a few days before. We were told to face one on the front and one on the rear. That didn't sound too good.

After daylight, a first lieutenant came along the ditch, and I said, "Sir, what is the deal? What kind of shape are we in?"

He laughed and said, "Everything is about normal. We are cut off as usual."

He didn't seem too worried about it. That helped some.

We went through that day. The artillery fire was fierce, and at 3:30 in the afternoon there as a whistle and an explosion on the hill in front of us. Out of the trench ran thirty Germans!

I told this boy, "Keep a watch to be sure none of them slip

around and come from the other direction."

There was one of our men a few yards down from us. He was known for being an expert with a B.A.R. (Browning Automatic Rifle). He, with a little help from us, made short work of the counterattack. It was strange, to me, how the Germans did the same thing each day at 3:30 P.M.

After daylight, there were eight or ten stragglers that didn't make the crossing. One was a guy named Bussey. He was Captain Lesser's messenger. He reported in and Captain Lesser said, "Bussey, why did you run off and leave me?"

Bussey said, "Captain, I was damned scared. I just couldn't help it."

Captain Lesser said, "I know just how you felt. If you had lied to me, I would have had you court-martialed."

Nothing was done to anyone who came over the next day. The soldier that I sent to the aid station never came back. You will find out about Pierce (yes, he is the fellow that was in trouble back at Camp Barkeley) a little later.

We had a cold, wet, and miserable night, and when the sun came out the next morning, my partner, the eighteen-year-old boy, and I decided to scrape some of the mud off of our blankets to see if we could get them dry. He was holding the blanket, and I was scraping when all of a sudden there was a loud explosion overhead. I jumped into the hole and grabbed my partner's hand to drag him in the ditch.

His hand went limp. I knew he was dead. A piece of shrapnel hit him high on his left chest. Because it was an overhead burst, it traveled down and went into his heart. I got a medic, but I knew there wasn't any use.

I didn't know his name, where he was from, or anything. He was a very religious man. His Sunday School teacher wrote him a letter and gave him his Sunday School lesson every week. He had it out and was working on it before he was killed. Not knowing what to do with it, I unbuttoned his coat and put it over his heart and buttoned it up again. Needless to say, that was a long and lonely night. (This incident was extremely painful to Leo and still is).

Germany
15 December 1944 Day **989**: 90th Infantry Division, 2nd Battalion, 359th, Company F

Dear Zell and all,

How does this find everyone? I am just fine. Things have been plenty rough since I wrote you the last letter. I am sending a clipping out of Stars and Stripes. I am sure that it is a brighter picture than I could paint of it anyway. It starts on one side and continued on the backside. The one main thing that I saw wrong, it sounds as if our Regiment had very little to do with the deal.

I got about 20 letters—the first mail in several days. It sure did help my morale.

News from home sounds good. I just pray that God will see to it we can all be home soon. I have done more praying than sleeping lately.

Write often,

Love to all.

Leo

Germany
16 December 1944-Day 890

Dear Folks,

This leaves me just fine. I had several letters from you and nearly everyone else this A.M. It was the first mail I had received in quite a while. I got twenty letters. It sure did make me feel better. A lot of them were sent to the hospital and to the replacement pool in England.

As far as I know, thins are about normal over here. You sure can't tell anything about it by being here. You wonders what everyone else is doing or if your outfit is the only one fighting.

I had a couple of letters from Delbert. He was home on furlough then. I don't know any news, so just keep your trust in God and keep praying—that is what I am doing.

Love to all,

Leo

INTO THE FRAY

Once more into the Fray...
Into the last good fight I'll ever know.
Live and Die on this day...
Live and die on this day...
--Jon Treloar

Just after dark, Captain Lesser came out and said that we were pulling back across the Saar. He said, "At 2:00 A.M. sharp, we will pass right here and be ready to tack on to the last man. If you miss him, you are on your own, so head for the river."

Everything went well. The river was within its banks, and we didn't get our feet wet. Getting back to my feet being wet, I guess that ice water was what they needed. The swelling went down, and I was faithful to put on a dry pair of socks every day (not clean but dry).

We left a holding force across the river until we got back across. I sure was glad to see all fourteen of them come back at about 4:30 A.M. We sent a patrol back over there three days later, and the Germans were still shelling that hill. From the number of good men that we lost over there, it seems like there could have been a better way.

We set up a lot of defensive measures in case the Germans

decided to cross over. We dug our mortars in to put our machine-guns on high ground. I saw this cabin out in the woods, and I thought that it would be great for the machine gunners to stay in when they weren't on a gun. So I walked over there and found that somebody had beaten us to it. I started back and someone was yelling, "Go back! Go back! You are in a minefield!!!"

I looked it over and said, "It isn't any further to walk on over as it is to turn back."

Besides, I just crossed over it once. I saw two dead horses that had been blown up from the bottom. I just thought, "A shell hit them."

It was a land mine. I don't know if I missed all of the mines or if it was just frozen enough to hold me up. I think that it was the Lord who was on my side again. What I should have done was to ask for one of the engineers to meet me with a mine detector. Later, the engineers took a truck full of mines out of that field.

The first night we were back across the river, this guy and I were in our sleeping bags on top of a mattress. There wasn't any cover. We had a fire going in the stove. Sometime during the night, he saw the fire in the stove flickering out on the floor. He lunged off of the bed (still zipped in his sleeping bag) to put the fire out. He thought he was still on the front line. I had to get him out of his sleeping bag.

Christmas

We got back in time for Christmas of 1944. It had been a long and hard year, but a lot of things had changed. How it was managed, I don't know, but we had turkey, dressing, and all of the works. We hadn't seen the cooks in a long time, but we found out that they had been carrying ammunition and rations across the river to us for several days. Old Red Allen said, "I dreaded watching the guys come through the line for fear that you wouldn't be there."

But, I was.

"I remember a Christmas when I was young (probably around 10) Leo told us about his Christmas "stroll" to Luxemburg. He didn't mention guns or gore just cold, walking in the dark and getting on and off trucks that were the only place they got to sleep. I ask my Dad about it on the way home. He said Leo was in the biggest battle war and that many men went to that battle hungry, frozen and fatigued. He said many men going to the fight knew they would probably die. But they knew they had to fight and win. I said, "Leo didn't think he was going die!"

Dad said, "No son, he knew he was going to die. But, by the grace of God he was wrong."

That was the day I thought maybe war wasn't like what Hollywood presented it." Terry Travis (Leo's Nephew)

Just after chow, there was a U.S.O. Show group who came up and used the temporary mess hall where we had just had dinner. They were called a "Jeep In Show." All they came up in was a jeep. One man who was the M.C. was about fifty years old, I guessed, and there were five girls who sang and danced for us. The M.C. said they were called

"The Five Pips and a Drip." He stated that they assigned the girls to him because they were perfectly safe.

They hadn't been gone but very few minutes, and we had started back to our house that we were occupying. Then, two German fighter planes came down the street and shot that building all to pieces and shot up the street and the buildings. They must have gotten word somehow, but it was a few minutes too late. No one got hurt.

I was one of the few that were called back to help train replacements for three days. Our colonel was in charge. He was a West Pointer and looked every inch of it. You could tell the West Point graduates by just looking at how straight their backs were and how straight they stood. I found out that he wasn't a bad lot. For a few minutes, one day he and I had a little shooting match to see who could keep a tin can roll with a clip of ammo in a 45 automatic pistol (nine shots). Then he asked each of us what we wanted to accomplish during the three days. When it came to my turn, I told him, "I want to teach the replacements how to stay alive for three days."

He asked, "What about after three days?"

I said, "Most of them will take care of themselves after that. During the first three days, they are liable to shoot you or their buddy."

One of the replacement soldiers asked me, "If you ask for the password and they don't come right out with it, you let them have it, right?"

I said, "If you want to shoot your best friend, that is a good place to start." I told him, "The Germans always wear long ankle length overcoats and we don't wear them at all. They wear hobnailed shoes. They have a shorter choppy step. There is also a difference in the helmets."

I told him and the rest of the group that I was tempted to shoot first and ask for the password later. This guy had had on an ankle-length raincoat and was slipping along like he didn't want to be seen.

I halted him, and he yelled, "Short Stuff! Short Stuff!!"

I startled him, and he couldn't come up with the password, but I knew his voice. Short Stuff was his nickname because he was short, and that is why his raincoat was ankle length.

He had some good news. We had gotten some tanks across the river. That was good news! After three days, we boarded trucks and headed north. We had taken a lot of replacements. A lot of them were eighteen years old with six weeks of training. Then, we got several from the 8th Air Force. Most of them were the undesirable soldiers. Many of them had faced court marshal. Some of them had been in the army for over three years, but you know they wouldn't send their very best men to us (which is what we needed). Germany

Germany
3 January 1945-Day 1008

Dear Zell and All,

How does this find everyone in 1945? I am okay, in fact, I have been a lot worse places lately. We had a lot of turkey Christmas and New

Years. We also had a good U.S.O show and movie. That really helps a guy when his nerves have been under a strain like our was for a while.

The news seems to be looking better than it did for a while. I got your v-letter last night. It was the one that you asked me to tell you if it or the air mail letter got here first. I never received the airmail yet, but the airmail usually gets here first, even if it doesn't, I had rather wait a day or two for the airmail.

I have received two Christmas packages--one from Iva and one from the folks, so yours should get here soon.

Everything has frozen up now. We have had snow on the ground for about two weeks now. That really is better than all the rain and the mud.

I guess everyone is about through with their work by now.

You asked me if we ever got a chance to go to church. Ye, we do sometimes. It all depends on what we are doing and where we are.

A day like this would be good to play 42 or shoot pool or go to a movie, but that is about all. You know, war waits for nobody.

Answer Soon,

Love to all

Leo

P.S. Tell Buddy, Dora, and the kids I got their Christmas card a couple of days ago. I was glad to know they still remember me.

Ardennes Offensive

(The Battle of the Bulge)

This was when the 90th Infantry Division got involved in the "Battle of the Bulge." It was just before Christmas Day when Patton's Third Army went up to liberate the 101st Airborne Division. Before the Allies could cross the Rhine River, they had to face the last stand German onslaught in the Ardennes Forest (formerly called the Black Forest because the trees were so thick that it was dark inside). Hitler planned a swift breakthrough to capture Antwerp and split the Allied Army in two. Field Marshal Walter Model directed the surprise assault under cover of heavy fog; 38 German divisions struck along a fifty-mile front on December 16th. Mechanized units overran several First Army positions.

Model's armies drove the Allies almost to the Meuse River and surrounded them in the southern Ardennes. Asked to surrender, Brigadier General Anthony C. McAuliffe of the 101st Airborne Division at Bastogne replied, "Nuts!" (The story of the NUTS! Reply By Kenneth J. McAuliffe, Jr. "The U.S. Army" December 12, 2012)

Third Army armored units pierced the German lines and relieved Bastogne. This attack in the Ardennes became known as the "Battle of the Bulge" because of the bulging shape of the battle lines. By early January, the Allies had recovered all the ground they had lost in the battle.

The 359th Regiment made our first attack on January 10th. Our job was to cut off the bulge. We captured a lot of men and equipment. The entire operation lasted over a month.

We went to Luxembourg. When we got there, we stayed in a farmer's big barn.

The horse right kicked good old Stanley Skay where the shrapnel had wounded him earlier. Stanley Kembloski was back with us. He was with us from the start. Also, O.D. Desenberg survived the Saar River ordeal. He was a previous replacement. He was 34 years old and had a few gray whiskers on his chin. We called him "Old Gray Beard."

We were riding in a truck convoy to make our first attack on the "Bulge." A young infantry replacement asked, "Sergeant Brown, I guess some of us will be killed before night?"

I told him, "Let's hope not--I've been here a long time, and it hasn't happened yet."

Within fifteen minutes we had reached our destination, unloaded from the trucks and moved forward on foot. After going about a quarter of a mile, I saw the young replacement lying beside the road, dead.

I went over to check him out and couldn't find any blood, body wounds, or shell holes in his clothing. The fact that he was barefooted further amazed me. (perhaps killed for his shoes?) It was many years before I finally quit seeing that poor, young man dead by the side of the road.

We had pretty well stopped the forward thrust of the bulge. Our leaders wanted to cut the bulge off and take a lot of men and equipment. The weather was bitter cold. It snowed all the way up to our destination. We found out the next morning that we had a new battalion commander. He was about forty years old, a West Point graduate, and about 6' 6" tall. He was the best officer I ever saw. He was with the lead company all the time. When a shell came in, and we all hit the dirt (in this case, we hit the snow), he never got down – never. If you hit the dirt belly down, you are supposed to increase your chance of survival by 70%.

We had shells coming, and I made a jump for a hole, and one guy and Desenberg was already in it. So, I was about half above ground level. Blood was coming out of my ear, and the whole side of my face had no feeling in it. But, I didn't get any shrapnel in me. We had three of our men killed nearby. Since that day, I have never had good hearing out of that ear.

After about half a mile, we got into some heavy woods where we found the company commander and the rest of the company. Beyond the heavy woods was a space, about three hundred yards of open ground where the snow had blown off of it. There were another two patches of heavy woods. Captain Lesser told me to take my platoon (we had a new second lieutenant, but he didn't say) to one of the patches of woods. Then, he assigned another platoon to the other patch (Captain Lesser also had a new first lieutenant as company executive officer).

We took off in the snow—a foot deep. As soon as we started, the Germans began firing at us with small arms. We were running and firing. We ran and made it to our woods (we were very lucky there was no enemy there). Shortly, the artillery started to fall, and we were getting tree bursts. We scattered to cut the loss. One of the new replacements got hit in the top of the head. I am sure his helmet stopped his instant death, but it looked really bad to me. We didn't have a first aid man (there was only one per company).

I told the lieutenant, "We need to get him to some help!"

He asked the men, "Any volunteers to get him back three hundred yards to the company and to help?"

No one volunteered. I told him, "I will take him to an aid man."

He was a nice eighteen-year-old boy. I felt like his dad rather than his platoon sergeant at 26 years of age. As soon as we started back, the German soldiers from the woods on our right started firing at us. So, we got on our bellies in the trail that we had just beat out coming over. He could crawl pretty well. The bullets were coming close, and by the time we were halfway across, he had to have help to get along by having me pushing him in front of me. I realized that our packs were sticking above the snow and that is what they were shooting at.

I took my knife, cut his straps, and slid his pack up on the snow, then worked out of mine and did the same with it. We moved on with them shooting at our packs. We finally made it, and I got him on a stretcher, then to a jeep, and finally on his way to an aid station. I hope that he made it, but I never knew and usually didn't.

This new company executive officer came over to where I was and said, "Go back over there! You only brought the boy across because you thought it was safer over here."

I answered and said, "I am going back to my platoon, and I will show you that I am not afraid!" And with that, I started back. Captain Lesser said, "No, you're not! There isn't anything you have to prove to anybody and get yourself killed for it. I need you here."

I said, "Well, I have to go get my pack anyway!"

Captain Lesser said, "Just settle down! You can get it just as well after dark."

When it was dark, Captain Lesser said, "Leo, you can go get your pack."

Our lieutenant and platoon leader turned me in for the Bronze Star for saving the boy's life. The order came back to me to be redone, and the company clerk came to me to repeat the report. There was no way I would write myself up for a medal. (I received the Bronze Star in October 1985, forty years after the war was over. Why it took so long to get, I don't know).

Night Attack

(Note to the reader: Leo's daughter and I have visited the location, traveled this route, and have been where these events of this night took place. Patton gave the order to take the objective before morning. It was not done, and there was no time like the present, so the 90th stepped up to the plate and in deep snow and subzero temperatures. Little could they know that they passed by 10,000 elite Nazi troops on either side of the narrow road who were so shocked by the Tough 'Ombre's fearless courage. The foxholes remain to this day in the darkness of the Ardennes.)

We came to where the company was. I thought we just might get some hot chow, but we didn't. I expected to be told to form a defensive line and dig in for the night. The new lieutenant colonel came up and said, "General Patton has just left, and he said we had to cut that bulge off before they got a bunch of men and equipment out."

He then continued, "I don't give a damn if you get up in the middle of the road and take that hill. It has got to be done!"

The lieutenant colonel said, "We will use the paved two-lane road and put a company out as flank guards on each side."

Our Company F would be the lead company, and Company E and Company G would be the flank guard. It was pretty heavily wooded. The road that we were on was on top. There wasn't too much snow, maybe 10 inches. Out where the flank companies were, there was a lot of snow, and there were gullies with even more snow. There were a few wire fences covered with snow. We were not making any time at all. We only had two or three miles to go.

The lieutenant colonel was walking in the middle of the road, but we were stopped all the time. E and G Company couldn't make any progress, and the colonel said, "Call them up here to the road, and we will do it General Patton's way!"

I found out later that we were taking a whole regiment of 3500 men, a tank battalion, four or five T.D.'s (tank destroyers) and some ambulances. We had done just fine until we nearly reached our objective. Two huge German tanks were sitting side by side, facing us. They started firing their machine guns. They had a lot of tracers in their ammunition. We hit the ground (rather, we hit the snow)! Those tracers were lighting things up like it was daylight. The lieutenant Colonel came along and then bent over slightly, and I could see by the tracers that the bullets were missing his helmet by a foot.

I was fifty feet from the tanks, but the scouts were within a few feet of them. It was very dark and freezing. Until the tanks started firing the tracers, we wondered why they didn't shoot the big tank guns. The lieutenant colonel said, "What is the trouble here?" Of course, he knew, so he said, "Pass the word to crawl back! I will call artillery on them!"

All the time, he was standing in the middle of the road. He called his radio man up (artillery observer) and stood in the road and directed artillery fire on the tanks. The shells never directly hit them, but I bet the shrapnel did, and that is why they took off.

Our objective was only about a hundred yards further on. There, we captured a full colonel and his staff, including an intelligence man. We had our intelligence man, and between him and the lieutenant colonel of ours, they convinced the Germans that the very best thing they could do was tell them everything they wanted to know. They found out that we had bypassed about ten thousand German troops and a lot of equipment, such as tanks.

We brought in more reinforcement for the rest of the night. We captured quite a few prisoners; however, many soldiers got away. Captain Lesser got the German colonel's dugout. It had a light and a heater. It was better than what he had been used to.

When daylight came, we were on a hill, overlooking a valley with a good road at its foot. Our lieutenant colonel was early. He had been observing a lot with his big field glasses. He went over to the tanker and said, "Boy, have I got good news for you. I can show you where the three German tanks are. You can shoot them just like you would shoot fish in a barrel."

The tanker got a little pale and asked, "Three, sir?"

The colonel said, "Pull over here, and I will show you."

The tanker still couldn't see the tanks. So, the Colonel got in the tank's seat and took a shot and missed. The tank gunner pushed him out of the way and took the next three shots. He hit all three German tanks. By this time, the American tanks had a much better gun on them compared to the one they started out with.

A lot of men and equipment were lost in this "Battle of the Bulge" we read and hear about. At this time, it hadn't gotten that name. It was terribly cold; we lost a lot of men to the enemy, and the weather also cost a lot of lives. It seemed better than the Saar River. At least here we were making some progress. We were on the move, and this gave us hope that this would eventually be over.

After the Germans lost this last big gamble, they too knew one day it would be over, and they knew who the winner was going to be. They caused us a lot of casualties after that, but I don't think they had the will to win. That is what it takes. If they had them the will, then I don't know why they didn't kill our lieutenant colonel standing in the middle of the road facing two German tanks. I don't. This past action took place on January 12, 1945.

I think that the Germans were completely surprised by how quickly we counter-attacked the bulge in their line. It is a big operation to move an army and all the tons and tons of equipment and supplies—fuel, ammunition, food—as it was done in inclement weather and on icy roads. The temperature was about twenty degrees below zero at night and about zero degrees in the daytime.

We lost a lot of good men in the "Battle of the Bulge," I think about as many to the weather as the enemy action. The war was in the process of winding down, but that didn't mean that the Germans wouldn't kill you if the chance came. There were some more rivers out there to cross and thousands of hills and towns to take. You never knew which ones the Germans would decide to defend. But, the "Bulge" was no more, and we turned eastward. That meant that Germany lay just ahead of us along with the Moselle River, Frankfurt, the Siegfried Line but a lot of pillboxes had to be taken but, the Siegfried Line was covered by overlapping firepower, mines, barbed wire fences, and dragon's teeth (concrete pillars crisscrossed at angles densely positioned along a line were used to prevent the movement of trucks and tanks).

Directly ahead of us was a little river to cross. I believe it was the Orr River. We were issued three days of K-rations as was the custom on a river crossing. We didn't pass for three days for some reason. No one checked to see if we still had three days of rations as we usually did.

We were on a hill overlooking the river waiting for other troops to cross. Shells started falling around and among us. Captain Lesser yelled down the hill to the river to warn us. He figured they had picked up our position. We went through a grove of small trees. A sergeant ten feet in front of me hit a booby trap mine about three feet off the ground. He yelled, and it knocked him down. He said, "Brown, help me! Don't go off and leave me."

I said, "I'll help you."

Of course, I figured if he hit one, there might be more there. I figured if I came the way he did, there wouldn't be any trip wires, or he would have set them off.

Captain Lesser said, "Leo, this is the medic's job. I have got to have you to lead the platoon."

He told the medic to unroll his pack and roll him up in a blanket and that he would send a jeep for them, which he did.

We continued on our way to cross the river when we found out where we were. We had already crossed the river on the ice and a couple of feet of snow. The captain told us to dig in. I was as tired and pooped out as I ever was unless I compare it to the night we crossed the Saar River.

About that time, more replacements were sent. After the sergeant had been wounded by the mine, I had four men, including myself. I told the new guys to form a line and dig in. One of them said, "Sergeant we don't have any entrenching tools (shovels)."

I said, "Eat your rations, and we will hand these down to you."

They didn't have any rations. I found out later that my three men didn't either. So, I shared mine with them. In a few minutes, one of the new boys came up to me and said, "Sergeant, when do we get some ammo?"

I replied, "You don't have any ammo?"

He said, "We have never had ammo!"

I went to Captain Lesser and said, "Sir, did you know these new men don't have any ammo?"

He said, "I'm just now finding that out!"

He took his radio, called regiment headquarters, and said, "I would like to have these replacements, but I can't use them as they are. We don't have any names, nor do we know anything about them. I will send them back to you, and when they are equipped and in the daylight, I would like to have them back."

The colonel at regiment headquarters said, "O.K.!"

A captain usually doesn't usually talk like that to a full colonel, but he got away with it that time.

About that time, the replacements had headed back, and our battalion commander, the lieutenant colonel, turned up and said, "I just found out that the town around a couple of miles from here, the one that we were to take, is not occupied."

We were going to get it before the Germans did. So, I did some fast pack-rolling. My platoon got the flank guard (it was bitter cold that night). We occupied the town without a problem with exception to one guard. When I got to town, I found I was the only one who had a pack and blanket. We had to send the men back to get blankets.

We found a hay barn with enough room to build a fire. The fire felt good, but we didn't have food. That made two days without food, and the twenty-degree weather was about to do us in.

The next morning, there was a lot of yelling that the Germans were coming. It turned out they were. We went out to meet them. They weren't very far, and we opened fire; the M-1 rifles fired one time, and the machine guns wouldn't fire at all. That stopped them, and they all dispersed because they didn't think the Americans were there. Our lieutenant colonel grabbed an artillery radio and went out and stood in the middle of the road and called artillery fire on them until the ones left had enough. I still don't know why someone didn't kill him, but they didn't.

The reason the M-1 rifle wouldn't fire but once was because the slot behind the bolt had blown full of snow, and when it fired, and the bolt came back, it mashed the snow into a ball of ice. There was no room to eject another cartridge. Oil will freeze at that temperature. We learned that you don't oil weapons at twenty degrees below zero. We had to clean the snow out of the M-1 rifle regularly.

That day, they got tanks across the river. I went and asked if they had anything to eat, and they gave me some C-rations. They had been frozen for two months, but we ate them, and finally, the kitchen truck arrived that night. Old Vernon (Red) Allen made pancakes as long as anyone would eat them. I ate nine pancakes and a serving of oatmeal.

The next day, we moved to the east and met the replacements we had turned down a few days before. We stayed in an occupied town that night. We had more hot chow. I felt so sorry for those replacements. I got them in one room and talked to them and told them that the weather had taken as many lives as the enemy.

I said, "I will never ask you to do anything that I can't do or haven't already done."

One reason that I felt so badly was they could see four of us when there should be fifty (which there hadn't been for a long time).

One of them spoke up and said, "Sergeant, you don't have to worry about us. Nearly all of us are from Texas. We won't give you any trouble."

His name was Henry Bell from Friona, Texas. Not all of them

were from Texas, but several were. One was Red Cadwell. He and his friend (I can't remember his name) were from Borger, Texas. We had a soldier from Kansas. He had two kids, and we called him "Pop Carson." Old Stanley Skay was back, and O.D. Desenberg was still with us.

We were back in Germany now, and the snow was turning to rain and melting a lot of the snow. It would freeze up at night. We were running into a lot of pillboxes. These had all been cleared before the Battle of the Bulge. A lot of them were back in enemy hands. Most of them were occupied, but the resistance was fairly light.

We figured most of them had pulled behind the Siegfried Line where all the "dragon's teeth," mines, and barbed-wire fences were. Captain Lesser notified us that he was sending several of us and all four platoon sergeants to rest camp for three days. That is what it was. We ate well and got a haircut and clean clothes. It rained all the time we were there, but it was safe, warm, and out of the weather. There was plenty to eat. What more could you want?

We returned at about twelve midnight. We were informed that we would make an attack on the Siegfried line at 2:00 A.M. My platoon was on an outpost half of a mile in front of where the rest were. There was an old tank (like the one we had in Normandy with an airplane engine in it that made a lot of noise).

It would go to the outpost and turn around and come back, really revving it up! When we left, the tank stayed. It was still going back and forth making a lot of noise. We had about one and a half miles to go where there was a little town that was split by the Siegfried line.

We went to all the houses (the were all vacant) and got all the beds, quilts, and everything we could find. As we walked across the "dragon's teeth" we put the mattresses ahead of us and covered them with quilts and blankets. This was to keep from setting off the land mines and to get over the barbed wire entanglement.

Just then, we crossed the all-talked-about Siegfried Line. Just to the right of where we crossed was a gate. It had a big concrete post on each side. They had huge notches in them with twenty-four-inch channel iron in them. They weighed several thousand pounds each.

There wasn't a single person defending the gate. Figuring the Germans is hard. We occupied the buildings to the side we were in now. They were mostly three-story reinforced concrete. They were designed to withstand a lot of artillery or bombs. Nothing more happened until daylight.

When the Germans discovered that we had crossed the line, they started throwing the artillery on us. The reinforced concrete houses were made to stand it. We found out that these good, strong houses were built on top of big pillboxes. When we threw a few phosphorous grenades down the vent pipes, they came out with their hands up.

We had one boy in our platoon slightly wounded after we crossed the line. That was our only casualty. We had heard about this Siegfried Line for two years. The prisoners wanted to know where all the "Panzers" (tanks) were. We told them that we only had one and that we didn't bring it.

They said, "We heard you all night bringing tanks in by the thousands (the one old tank we were driving back and forth all night)!"

The G Company commander (he came over as our platoon leader and took over G Company after the first day in Normandy) out-maneuvered seventy S.S. troops with a big self-propelled gun and captured them with a few of our soldiers. I don't remember the details, but he received the Congressional Medal of Honor for it. I was glad for him because it had been a long fight, and he hadn't missed a day.

When the trucks, tanks, and other vehicles tried to move up, they ran into a lot of mines. The ground was frozen solid enough that we foot soldiers never set them off. Those mines, I guess, were probably what the Germans depended on for protection.

An engineer captain came up and blew the heavy steel channel iron in two. A tank dozer pushed it out of the way and opened the gate to Germany.

The best fortifications in the world are no better than the people who manned them. The rain continued. Foxholes would seep fully of water, and we were wet including our blankets. There were rolls of blankets in a trailer behind the kitchen truck, and they were wet too! Captain Lesser told the supply sergeant and his helper that he just had to get the blankets dried out and the supply sergeant said, "I would like to know how I am supposed to get blankets dried while it is raining!"

The captain answered, "If I were a supply sergeant and wanted to remain one, I would take some clothesline and stretch it back and forth in one of these big houses and build a fire in the stove. I would dry these blankets so that I could be a supply sergeant the following day."

We had half of our blankets dried that day and the remaining half on the following day.

Due to time overseas and all the wet weather, no one in my platoon had a watch that would run. We operated on time a lot, and we always stood guard. We found an old alarm clock. One day we were making an attack on some houses at the edge of town. We were to be as quiet as possible. Guess what happened? The alarm went off in this boy's pack! That night I told the supply sergeant that I had to have a watch. Every evening, the company commander's messenger had a GI watch, but he said, "We don't have any watches. It's just that simple!"

I figured he was saving them to take home or was sending them home. So I said, "You have one, and you don't need it." I continued, "So if you don't get me a watch in three days, I will just take yours until you do!"

He said, "How do you think you will get mine?"

I told him, "I haven't got that figured out yet, but the only thing I am sure of is when it is over, I will have the watch!"

I walked off, and a couple of days later, he came to me and handed me a brand new G.I. watch. I am sure he had it all the time.

If we had still had Jack Sharman, the supply sergeant that was overseas with us, this incident wouldn't have happened. Jack was very diligent to keep authorized supplies distributed out to the people allowed to have them.

One night Captain Lesser called me up to his CP (Command Post) and wanted me to take the 3rd platoon. They had more men than the 4th. He said, "I am going to recommend a couple of the platoon sergeants for a battlefield commission."

I told him, "I'm not interested in a commission." (That was a big mistake.) I continued, "I like the boys and would feel like a traitor if I left them."

He said, "I see how you feel and I won't try to force anything on you."

By the time we got back to the Moselle River, I was back to sixteen men. Captain Lesser and all the platoon leaders went to look at the river. It was within its banks. It was a big old river. It was at least a half mile wide. It carried commercial steamships. There were two artillery observers there. There were big hills on both sides of the river. We were going to cross at 4:00 A.M. of the following day on assault boats. It was quite a river to cross with a sixteen-foot boat with paddles.

The map showed a paved road just a little above the water's edge. The captain told us not to tell the men that night because some of them wouldn't sleep if they knew what was coming up. It didn't look good to me even if there wasn't an enemy to worry about. We were going down in single file to the river. The word came back (as it often did) not to bunch up on the footbridge. Boy, that was good news, but when we got there, there was a stack of boats and my platoon split up. Eight of us each got in a boat with two engineers per boat, one in front and one in back. They did all the steering. I took half, and Stanley Skay took the other half. I said, "I will see you on the other side. If I don't see you in thirty minutes, move on inland and try to find the rest of the company and I will do the same."

We got a paddle and pushed off. The water was swift, and when we got out in the middle, we hit a whirlpool. The boat spun like a top. I thought, "This is it!"

How, I don't know, but the engineers pulled us out of it. We paddled onto shore and dragged the boat out of the water. There was the paved road, and old Stanley Skay and his men were just dragging their boat out. They had the same experience.

We went up the steep hill and found the rest of the company. We had some counter attacks, but the artillery broke them up. I suppose the artillery was still across the river. I am sure we had the best artillery in the world and plenty of ammo. We met resistance, but it didn't last long.

As the supply lines began to get longer, some things began to get harder to get—food and cigarettes were two of them. We were getting plenty of K-rations but no hot chow. I told my old friend Vernon T. Allen (the cook), "You guys are just going for a sightseeing tour. You never cook us anything."

He said, "You get me some food, and I will cook it!"

We were sending out patrols regularly to see that we weren't passing up the enemy. One morning, it was our job to take out a patrol. As we came back to the road, about half of a mile behind the company and the kitchen truck, there was a little stream and the ground dropped off about six feet. The Germans had taken the 16-inch pipe and laid it straight out in the direction of the flow. At the end of the pipe was a little water wheel that turned a gristmill, and beside it was about 1500 pounds of whole-wheat flour.

After we got our flour and were heading back to the kitchen truck, we met a few German civilians. As we were also gathering military information, we stopped the travelers to speak with them. I could speak a little bit of German, so I was asking the questions. I knew that "Vohin gehen Sie?" meant "Where are you going?"

I didn't care where they were going. I wanted to know where they had been so I didn't know how to ask that. So I made up a word and asked, "Vohin been Sie?"

Of course, they didn't know what I asked and began to laugh. Their laughter was at my expense!

We got about three bushels and carried them to the kitchen truck. Captain Lesser was there and said, "Be sure to make us some pancakes."

Our good lieutenant colonel drove up, got out of the vehicle, and looked around, then he asked Captain Lesser, "How did you come up with pancakes while everyone else is eating K-rations?" (I was close enough to hear what was being said)

The captain raised up and pointed straight at me and said, "That's Sergeant Brown, and he will tell you how we got the flour for the pancakes."

So he came over, and I started coming to attention, and he said, "Keep your seat and enjoy your pancakes and tell me how and where you got flour while you were on patrol."

I told him, and he said, "Do you think that was the thing to do?"

I said, "I think it was because we have been short on rations for a good while and I thought some pancakes would perk us up."

He said, "I agree with you on that but what I disagree on is here is one company eating pancakes and the other three are eating K-rations. That I don't like." He continued, "How much flour did you find?"

I told him, "About 1500 pounds."

He said, "Do you think that you could get a jeep and a couple of boys and get enough so that the other companies could have pancakes?"

We did just that, so we all had pancakes, and everything worked out OK. I found that the higher-ranking officers were the easiest to deal with.

Things were moving faster now. Heavy columns of German vehicles were moving east, and the American fighter-bombers (mainly

P-47s) were playing havoc with them. A lot of these vehicles moved all night and hid during the daytime, but the American bombers found a lot of them. The trains were also targets. There wasn't a lot of room for them to hide except in some of the tunnels.

GERMANY AND THE CZECH

We passed through a town by the name of Merkers. It was the German equivalent to Fort Knox. It had Germany's gold supply and a lot of stolen art and other goods from other countries. It was moved and stored under guard. The gold and other supplies were stored underground in a salt mine. I am sure they got their hundred tons of gold bullion (I can only think how they would have treated us if things had been reversed).

Sergeant John Davis came back to the company. He was the third platoon sergeant. I was glad I didn't take the assignment Captain Lesser offered to me earlier. Then, the bad news came. We got two new second lieutenants. The one I got was about 45 years old and was transferred to us out of the antiaircraft unit. These kind were always bad news. Captain Lesser had company headquarters in one room of a house. The next room was the supply room. I was in the supply room when he was talking to the new lieutenants. First, he talked to the one that was going to the third platoon. Then he got to mine, and I am sure he didn't like what he saw. Then he said, "I am going to send you to the 4th Platoon, and that is Sergeant Brown's platoon. He has been running it for a long time, and he still is. You go and follow along and you might learn something, since he has been here since D-Day (he hadn't).

Don't you go there and give him a lot of trouble because if you do, you will have me to deal with!"

Captain Lesser had no idea that I was in the next room and had heard him. The lieutenant came to our platoon and tried to boss me around. It didn't work out at all. That is when I realized that I should have accepted the commission.

With us getting the second lieutenants, we lost our good battalion commander, the lieutenant colonel. He was transferred to Regiment Headquarters, G2 (intelligence). A new major replaced him. The next day, he called me over (he and Captain Lesser) where they were talking. They had maps laid out. He told us, "About one and a half miles from here is a railroad tunnel with six train engines in it. I want your platoon to go and seal it up."

It sounded like a pretty big order for one platoon of sixteen men. He assured me that it would be a snap. I asked him, "If we get in trouble, will we have artillery support?"

He replied, "All you want."

He gave me a map, and it had circles drawn with a red pencil with numbers called checkpoints. He continued, "All you have to do is tell me what checkpoints you want to be shelled, and we'll be ready to tear the timber down!"

I reminded him that we didn't have a radio, so he gave me Captain Lesser's. I also asked for a medical corpsman. We started toward our objective and were walking down a road. Red Cadwell (from Borger, Texas) and Henry Abeyta were leading (he was a Mexican boy, a good soldier). Desenberg was just behind them. All of a sudden, a lot of gunfire broke out. There was a German up on a bank with a machine-gun. Desenberg was too close for him to shoot him with his machine-gun. So, the German got out his pistol and intended to shoot Desenberg in the top of the head but cut a bullet groove down the back of his shoulder instead.

Desenberg underwent a lot of "razzing" regarding the fact that he got shot in the back while he was advancing. When he got back to the States, it was two years before he and his wife had another child. He had to take more "razzing" with questions like, "Now where did you say you got shot?"

We had one of our men killed, and three others wounded. I called the major and requested the ambulance and artillery fire. He told me to pull off the road, move forward, and they would send stretchers and medics to carry the wounded. We made another 150 yards and got in more trouble. Red Cadwell and a soldier named Miller were killed, and we still didn't get any artillery fire.

The little eighteen-year-old by from Greely, Colorado, was beside Red when he was killed. Henry Abeyta, we called him "Pancho," yelled, "Sergeant Brown! Sergeant Brown! Sergeant Brown! What should we do?"

I told him, "Stay behind the stump and get ready to run when I give the word."

I gave the word and told him, "Run!"

He took off as fast as he could and jumped into the ditch with me. The bullets from the German's machine pistols stirred up the dirt right behind Pancho, but he didn't get a scratch!

Captain Lesser came on the radio and told us to stay where we were, and he would bring the rest of the company and help us. He said that this was just a pocket of the enemy that was bypassed. He told me the artillery had been moved up ahead that morning. In other words, the new major lied to us. There wasn't any artillery to call on. As soon as it got dark, the Germans took off, and we blocked the tunnel, but we missed our fearless lieutenant. We found him huddled behind a rock, firing his rifle straight up in the air.

All this was such a shame. These were the last people that we had in our company that got killed. With the right kind of leadership, it wouldn't have happened. There should have been a huge artillery barrage before we went in there. If we had had artillery fire or one tank, we could have taken the tunnel without a casualty. I am sure.

The German soldiers, by now, were only defending towns and important road junctions. It was pretty well-established that the Germans were leaving. There would be a white sheet hanging out of the church steeple that meant the town had surrendered.

Some units of armor and reconnaissance had already reached

the Rhine River. We were approaching the city of Mainz; it was a large city with factories in it. We received word that the city wouldn't be defended, but it was. We had over half of it occupied when they decided to defend the factories.

We were supposed to cross the Rhine River that night, but we had to clear the rest of the town. There was an artillery observer who came to us before dark and said, "I am going to wreck that factory with eight-inch artillery."

He did!

Captain Lesser then sent for us to come to his C.P. and get our map and orders for the next day. I got up, took my steel helmet, pulled my 45 automatic pistol out of its holster and stuck it in my belt and started for the door. The lieutenant put on his pack and everything and told me to do the same. I said, "We are just going down the street a couple of blocks!"

He gave me an argument. Captain Lesser had given me the instructions and map, and that didn't suit the lieutenant too well. When we left, he turned the wrong direction. I stopped him and said, "It isn't but half a mile to enemy territory. If that is what you want, go ahead."

He argued with me, but not too hard. When we got the rest of the town cleared out, the 5th Division had a bridge across the Rhine, and we crossed it on the run (on two bridges). There were about 35 German fighter planes around, but they didn't give us any trouble. There was a little town about half of a mile from the river, and this was new to us, but their police force decided to defend it. They didn't take things too seriously, so none of them were hurt.

There was a real level field with some green feed growing about six inches high. Shortly after we got there, a P-47 came in for a belly landing. The pilot jumped out with his 45 automatic pistols in his hand. He sure looked relieved when I hollered and said, "You won't need that!"

We headed in a southeast direction. We were guiding on the Czech border. We were sending out a lot of patrols. One morning, my platoon went out for a long distance. The grass was tall, and there were not many houses. We came to this house, and Stanley Skay started talking to the occupant. He and another soldier were waving their arms and pointing (Stanley was Polish and could understand most of the language). Pretty soon, Stanley said, "Let's go in the house. They are going to fix breakfast for us."

Boy, they did!

The woman had a big old cast iron kitchen range, and she cooked regular sausage, big old biscuits, white gravy, and regular butter. I had a jar of instant coffee. They hadn't had coffee for ages. We all left happy (we were the first Americans they had seen). When we left, I asked Stanley, "What did you tell them?"

He said, "I told them that we had been cut off and almost captured and had to hide out for three days without anything to eat."

I asked, "Did you explain how we all had a good shave?"

He replied, "Why worry? You got your belly full didn't you?"

“At 1:30 pm on May 4, the 90th Infantry Division headquarters, in anticipation of the formal surrender, had sent out orders that the battle-hardened, famous 11th Panzer Division assemble in two areas. Each one of them was to have a water point established in it, a task to be accomplished by the 204th Engineer Battalion. It was estimated that the 11th Panzer Division would be logistically self-sustaining for four days. The division’s elements would take two routes out of Czechoslovakia, one through the border town of Rittsteig, the other through Vseruby where the negotiations had taken place. The routes would terminate in two assembly areas, designated “A” and “B,” one on either side of the German town of Koetzting. The 2nd Battalion of the 359th had responsibility for Area “A” to the west of Koetzting, while the 3rd Battalion was charged with Area “B.” (Brig. General E. Raymond Bell, Jr.)

We met the first lieutenant from intelligence (G2). They had heard about the 11th Panzer Division out in front of us that was about to attack. We had noticed the lieutenant before we left so we would know him when he came back. He badly needed a haircut and shave. He had an old Russian uniform on that needed some patching and had a stick with a bundle tied to it. He looked like anything but a G2 officer. He said, "I just wander around and would eat in the enemy chow line if they let me. Then I just listen."

He continued, "They have plenty of men, equipment, ammo, but they lack fuel."

Sure enough, in a couple of days, there was a German officer with a letter from the General wanting to surrender the entire division. They worked out an agreement. We would line up on each side of the road, and they would drive in about half of a mile with vehicles and we would disarm them. Then another bunch would pull in. It took until dark to get this done.

We allowed the officers to keep their sidearm (pistols) for their protection from their troops. Their officers were in charge of the move, and there were two Americans in every other vehicle to see that we were getting to where we were going.

It was pouring down rain, as usual. We had already taken the wrong fork in the road and unhooked a big gun and pushed it into a creek and turned around. When we got back to the road, this German officer stepped out in front of us and tried to stop us. The driver of this old car tried to run him down. He jumped out of their way and tried to get his pistol out. I was sitting directly behind the driver, so I rolled the glass down and stuck my M-1 rifle out and pulled the bolt back and let it "fly to." That stopped the fast draw. I don't know what their problem was, but that driver intended to run him down.

We went on for another mile, and everything came to a stop; we didn't move until after daylight. Some American officers arrived and took over. The day was May 3, 1945. There were nine thousand fifty prisoners, seven hundred or more tanks, guns, half-tracks, etc.

The next day we were to break through the pass to Prague, Czechoslovakia, and let the 4th Armored Division go in and liberate Prague. The Germans were giving us a hard time. We had just got new tanks with 90-mm guns on them, but they never got to fire a shot. A bunch of people came running down the road yelling that the war was over! We got word over the radio to not move forward but to sit still. After a while, word came for us to go to a little village and wait. Before night, we got word that it would all be over on May 8, 1945.

Germany
Co. F 359th Inf., A.P.O.-90
4 December1944- **Day 978**

Dear Allen,

How does this find the old man? I am just fine.

It is just daylight and I had a stack of hotcakes for breakfast. We are in a little German village. It is just a ghost town. Most of the houses are torn down. There aren't any Germans here. They all took off. But, that isn't always true. Sometimes a lot of the civilians stay. They seem neither friendly nor hostile. It is much different than the French towns. If there was a German soldier there and they knew it there would be a fight.

The French would be in their basements. If the Germans pulled out there would be a lot of them out to meet us when we got there. You could always tell what to expect that way.

I had a letter from Zell. She said you were finished combining grain. Were the crops good this year? She you were also about to sell your north half-section. The price didn't sound bad to me.

News is scarce so write soon,

As ever,

Leo

1945

Luxembourg
28 January 1945 – **Day 1032**
90th Infantry, 2nd Battalion, 359th, Company F

Dear Zell and All,

I got a letter from you a couple of days ago. I was real proud to hear from you. I am in a room with a stove now. It is really okay but I don't guess that it will last long. It is the first fire I have seen for a long time.

It sounded like Allen got a good price for the old regular.

I had a letter from Ray and Lehman a few days ago. They were both talking about hot weather. This is about the coldest weather and the most snow I have ever seen.

I really like the way those Russians are headed to Berlin. Don't you?

You asked if I got the gloves the folks sent to me. I haven't yet but probably will soon.

Love to all.

Leo

P.S. How does Allen like his A.C.? (Allis-Chalmers tractor)

Paris, France
28 February 1945- **Day 1064** (VE day-69 days)
Leo H Brown, 90th Infantry, 2nd Battalion, 359th, Company F

Dear Zell and All,

How is everyone? I am just fine and I have had a wonderful time in Paris. It is quite a city. I sent Don a toy jeep I bought here.

I wish I could have bought something for everyone but things are so awfully high. Money goes like water.

It would take a week to tell all about Paris, so I won't start.

Write soon,

Love to all,

Leo

Germany
11 March 1945- **Day 1076**
Leo H Brown, 90th Infantry, 2nd Battalion, 359th, Company F

Dear Zell and All,

I am finally getting around to writing you. Things have been going pretty fast to write very much lately.

I am making out just find. The weather has warmed up a lot. It rains a lot and snows some, but it melts off fast now.

I had a letter from Delbert a few days ago. He was in France then. I judge he is with that new Army over here (just my guess).

You asked me, which was the roughest when we were moving fast, or when we made a few hundred yards, or made a few miles. I have always seen it as this, when we sit still it isn't so bad. The worst times are when you can only make a few hundred yards. They (the enemy) know where you are and you have to be in the open and they can shell you all the time! When we move fast, they have to pull their artillery back and half the time, they don't know where we are.

We have to keep moving and I do a lot of walking and all that stuff but other than that it isn't near as bad as moving slowly. When we move fast, everyone's morale is high and the enemy's is low. He doesn't know if he is cut off or not.

Another guy and I were looking around through a village a few days ago and we didn't expect to find any enemy. Three walked out of a house with their hands up. They were glad to be prisoners. The war was over for them. They had no hope at all to win. They said it would be over in a month (I hope the soldier who told me this is right). He said that Germany had no fuel. The one I talked to was a pilot until two months ago and he said they had no gas so he was put in the infantry. He spoke fair English. He said Germany would never surrender. He said we would make a big breakthrough and occupy all the country and it would all be over. I think he was right. It's hard to tell.

Write soon,

Love to all,

Leo

27 March 1945-Day 1021
Germany

Dear Sister and All

I have been trying to get time to write you a letter from several days now but have been on the move too much. I got the cake you sent me. We have just finished taking the city of Mainz, Germany. Maybe you read about it. It was in the Stars and Stripes. They made it worse than it was. Anyway, the cake was swell.

I haven't had an mail for a few days. I have had bout three letters from Delbert since he as been over here.

How is Allen and his new tractor coming along? Okay I guess?

Things have been going to fast to believe for the past month. We sure didn't get any rest and I hope the "Jerries" Germans don't.

You got the wrong idea about the pillbox I was in a few months before it was blown up. It was our engineers that blew it up, not the English. I haven't been around any of them since I have been here. We blew them up because if the Germans ever make another drive, they couldn't use them like they did before.

I had a letter from Cordia. She said that she and Art quit. I figured that out anyway from what I know about him. I don't think anyone could live with him.

News is scarce so answer soon

Oh yeah, I got another stripe the other day.

Love to all,

Leo

VICTORY IN EUROPE!

Victory in Europe Day, generally known as V-E Day, VE Day or simply V Day was the public holiday celebrated on 8 May 1945 (7 May in Commonwealth realms[citation needed]) to mark the formal acceptance by the Allies of World War II of Nazi Germany's unconditional surrender of its armed forces. It thus marked the end of World War II in Europe.

The term VE Day existed as early as September 1944,in anticipation of victory. On 30 April 1945, Adolf Hitler, the Nazi leader, committed suicide during the Battle of Berlin. Germany's surrender, therefore, was authorized by his successor, Reichspräsident Karl Dönitz. The act of military surrender was signed on 7 May in Reims, France and on 8 May in Berlin, Germany.

Somewhere in the Czech
11 May 1945- **Day 1136** (VE day +3 days)
Leo H Brown, 90th Infantry, 2nd Battalion, 359th, Company F

Dear Zell and all,

I received your package with the cookies and toilet articles last night. It came in just right since V.E. Day. Maybe we won't have to move on foot and have to carry everything we own in our pockets—I hope.

The old 90th was till attacking when it was over. You know how it was, so many rumors and long drawn out affairs. It was just over and everyone said, "I thought so" or "I knew it would happen" or "It was plain it couldn't last long."

We (the 359th Inf.) captured the whole 11th Panzer Division. They were nearly full strength on men and the right age. It was Germany's best Panzer Division. It was the one that made the breakthrough in Belgium. Their equipment was worn out and shot up but they still had plenty of ammo and four week's rations with them.

The 90th Division has captured 83,000 prisoners and has 318 days of combat time. What is in store for us is anyone's guess. We should get a break but I don't know and don't intend to worry about it.

You can send me some stuff now and I hope I have time to use it. A bottle of ink, some kind of writing paper, bottle of Vaseline hair oil, a couple of packets of double edged razor blades, a couple of magazines, True Romance or Westerns, quite a variety I would say.

It takes about three weeks to hear from Delbert. I can't figure why. I hope to see him but I doubt it.

I hope you have had a good rain before now.

Did the people in the states celebrate a lot when the news came that V.E. Day had come. I hope I know more to write next time.

Love to all,

Leo

Leo H Brown VE Day-**Day 1133**
8 May 1945

Carry On News Letter
Combat Team 359 APO 90
"This Issue May be Mailed Home"

"ENGRAVED IN OUR HEARTS FOR ALL TIME"
Today is Victory in Europe Day!

THE OFFICIAL ANNOUNCEMENT WILL BE BROADCAST AT 13 HOURS

GMT (1500 DBST). CONGRATULATIONS HAVE BEEN SENT TO GENERAL EISENHOWER ON THE SUCCESS OF THE ALLIED ARMIES. FIRST NEWS OF GERMANY'S CAPITULATION CAME FROM AN AGENCY CORRESPONDENT YESTERDAY AFTERNOON. IT SAID THE SURRENDER WAS MADE AT 0241 HOURS GMT (0441 DBST), YESTERDAY, IN THE SCHOOLHOUSE AT REIMS, GENERAL EISENHOWER'S HEADQUARTERS. THE SURRENDER WAS SIGNED FOR GERMANY BY COLONEL GENERAL GUSTAV JODL, THE NEW CHIEF OF STAFF OF THE GERMAN ARMY. LIEUTENANT GENERAL BEDELL SMITH SIGNED FOR GENERAL EISENHOWER. MAJOR GENERAL SUOSLOPAROV SIGNED FOR RUSSIA, AND GENERAL FRANCOIS SEVEZ SIGNED FOR FRANCE. THE GERMANS WERE REPEATEDLY ASKED IF THEY UNDERSTOOD THE SIGNIFICANCE OF THE TERMS. THEY REPLIED, "YES" AND SAID THEY WOULD BE CARRIED OUT BY GERMANY. According to the Czechoslovak Patriot radio, the Germans fighting in Prague have now agreed to unconditional surrender terms. The patriot radio said American tank spearheads were within seven miles of Prague. Russian forces driving west have captured four more towns within 120 miles of Prague.

This is the exact wording of the 359th News letter, "Carry On."

Salzbach, Germany
18 May 1945- **Day 1143** (VE day+10 days)
Leo H Brown, 90th Infantry, 2nd Battalion, 359th, Company F

Dear Zell and All,

Well I am back in Germany now. We were in Czech for a while. The people are friendly there. You could hardly walk for the flowers they threw at us. When we left they had the band playing. I wish I could speak the language they use there. It is a mixture of Czech, German, Russian, Polish, and I don't know what else.

Zell I am stationed in Salzbach now. You would have to have a small map to find it.

I wrote and told Delbert where I was. I hope he can come see me. I will try to see him if I can find out where he is. I don't know what chance there will be on either.

I have been about sick with one of those summer colds the last three or four days. I am going to go on a sick call tomorrow for the first time since I left the hospital.

I had a letter from Momma tonight. She said it was still dry there. I hope you have rain before now.

I hope they get us some movies or something fixed up for entertainment now that the war is over.

How did the people there take V.E. Day? It was good news to us because we were still attacking. It wasn't a surprise because we heard the rumors like you did. We knew it would be over anytime.

Write soon,

Love to all,

Leo

Salzbach, Germany
June 7, 1945 **Day 1163**: 90th Infantry Division, 2nd Battalion, 359th, Company F: D-Day +366

Dear Zell and All,

I got your letter of May 31st. It must really be dry there by now if it hasn't rained before now. It has begun to get late enough to be serious now. We have had several rains lately, regular old thunderstorms like back home.

I dug my first foxhole in France a year ago tonight. I will never forget that night a year ago and that morning when our ship sank (Suzan B.

Anthony) and saw all that stuff on the beach. There were four planes that went down. I am not kidding you! I aged 10 years that night and it wasn't bad at all after I learned better.

I heard from Delbert several times since the war was over. He is going to come up here I think. I am between Amberg and Nuremberg in a town named Salzbach. He is near Munich.

Boy, I really sweated it out the last two or three weeks. I know the war would end anytime, but a guy is just as dead if he gets killed the last day as the first, you know. Things look pretty dark to me. I read in the Army Times tonight where we were going to the Pacific with the 3rd Army. Delbert said the 20 A.D.is in the 3rd Army now, but it isn't mentioned here. I will send you the clipping. If you figure anything good out of it let me know. This was the first news like that we have had.

I haven't got the stuff you sent but it should arrive in a couple of weeks. Maybe I would need it by the then, who knows, but I tell you this, the next time anyone tells you that boys over here get everything, you just tell them they are crazy as hell for me! I would just like to go into one of those stores where they charge their prices and give you that old stuff and say that the good stuff is all going overseas! My feet are on the ground and I can't get shoes. I have turned in to supply for three months. We have one suit of clothes—the ones we fought in. We were going to get clothes, shoes, and balls and things to play with and recreational equipment and we haven't even have enough to eat and the boys can't get half enough cigarettes to smoke and I am getting disgusted!

The reason we didn't know things when we were fighting was we didn't have time to use it and couldn't carry it and we captured enough things to get by okay but those looting days are over.

No, Delbert didn't tell me about his Ford. I have seen a time when we had enough cars and motorcycles to haul the whole company. But you can't take anything you can't carry in the Infantry so we would have to ditch them when we move.

Zell, I hate to send this letter but I guess that is just the way I feel. You see these rear echelon guys with all the stuff then they tell us to go draw clothes so I line up about 20 men back. When I got up there I got two handkerchiefs and a raincoat that I didn't need. Then the mail came in and I read this clipping. It doesn't help your morale any.

Maybe I can do better next time.

Love to All,

Leo

Germany
14 June 1945- **Day 1170** (VE+37)
Leo H Brown, 90th Infantry, 2nd Battalion, 359th, Company F

Dear Zell and All,

How does this find everyone? I hope you don't still have the dry weather blues. I hope you have had a good rain before now. We really get our share here.

Delbert came up to see me the 10th. He just got to stay 24 hours but it was worth a million dollars to us. I really was glad to see him! It was pouring down rain when he left. I gave him my raincoat but I expect he got wet if he didn't catch a ride after I left him. I sure hope he got back on time. I kinda felt homesick after he left. We talked about home and old times. If I were to get another pass to Paris or some place, I am just going to go to see him. I have seen all the ETO (Europe) that I want.

From the way he talked, I wouldn't doubt if he isn't back to the States before I am (just a guess). There are a lot of rumors going around but I can't see that we are making any preparations (to invade Japan) to go anywhere soon. But, things can change in a hurry, you know. I don't know what to expect. The Army Times put out that the 3rd Army was going to the Pacific. Then Stars and Stripes said occupation. Then here comes the Portales News Tribune saying the 9th Army is occupation. If you can figure anything out of this, then you have beaten me. You never know who will be in that "certain" Army anyway. I would nearly bet that if General Patton goes to the Pacific the 90th Infantry Division and the 4th Armored Division goes with him. These are his main spearhead Divisions.

Zell, the Portales paper sends me as many as three issues of the same date. I don't know why. You can tell them that one is enough. The war isn't over in the E.T.O. (European Theater of Operations) in the Portales paper I get.

If you haven't already done so, can you send me a bottle of Jerry's Hair Oil? It is impossible to get any kind over here. Also, send me a set of dominoes. I have a lot of spare time now. You can ask Delbert what I do. That is about nothing when you get it added up--a lot of paperwork and not much else. The 1st Sergeant says he is getting discharged. They want me to take his job. It is a lot more work here, but in combat it is a much better job than Platoon Sergeant. I don't know what to do about it, in fact he hasn't gotten out and I am not sure he will. There are a lot, or I will say several guys due to go home that have not yet gone.

I had a chance to get a battlefield commission the first of last March as a 2nd Lieutenant. I figured I would have enough points to get out as an enlisted man but not as an officer, so I didn't take it. But, I don't have enough either way. I am not sorry that I did not take the commission.

Have many of the boys around there come home from over here yet? I had a letter from big Lars Todd the other day. He said he would be home soon and he thought he would be discharged (on what basis I do not know). He is in the 95th Infantry Division. They are going to C.B.I. (China Burma India).

Write Soon,

Love to All,

Leo

Germany
27 June 1945- **Day 1183** (VE+50 days)
Leo H Brown, 90th Infantry, 2nd Battalion, 359th, Company F

Dear Zell and All,

I got your letter of June 13. I was sorry to hear that it is still dry. We really had a lot of rain last night and today.

I was down to see Delbert last week. We had a nice visit and had nearly 48 hours together. I think he will be back in the States fairly soon.

We saw Price Gandy while we were together. We had a nice visit. Lowell Harp was gone. He is in Paris. Price said that he and Lowell were coming to see me. I hope they do. I hope Delbert gets to come see me before he leaves but I kinda doubt it. You can't tell about this Army. He might be in longer than he thinks.

I guess it is definite that we will go to the States and then to the Pacific, but I don't know when. The rumor is all the way from 15 July to 1 July, so you see, I don't have any idea.

We got two more battle stars today-makes me 4 now and I am still short 9 points. It looks like if you don't have enough, you may as well not have any. They never have set any difference of a man with 84 and one with 10. I hope they do.

I have changed my mind about the occupation. I would gladly take that instead of the Pacific. I have seen enough war to do me for a long time!

It seems like mail is slower since the war is over than it was before. I can't figure.

So far I haven't received any packages mailed since V.E. Day, I ought to get some soon, though.

The 1st Sergeant is getting out of the Army. They want me to take his place. I have until tomorrow night to make up my mind. There is a real headache with the job. I don't really know what to do.

Write Soon,

Love to all,

Leo

P.S. I saw a really good U.S.O. show last night, "Anything Goes" and it was good.

Germany
4 July 1945- **Day 1190** (VE+57 days)
Leo H Brown, 90th Infantry, 2nd Battalion, 359th, Company F

Dear Zell and All,

How does this find all of you? I am just fine. This is much better than a year ago today (Leo was wounded in France and men standing shoulder to shoulder with him were either killed or wounded when a tree burst exploded above them while they were loading ammunition).

We had a parade this morning. I was the color guard for the flag during the parade. That is supposed to be an honor. My right arm was so tired from carrying a rifle at the right shoulder arms, I couldn't straighten it out for about an hour after the parade was over.

All the German men have to stop walking and remove their hat when the American flag comes by. They sure don't seem happy to do so, but most of them do and the others do with a little coaxing.

I received the package with the airmail tablets, hair oil, magazines, etc. They were just what I needed. I had a letter from Delbert yesterday. It was about 10 days old. He said they had turned in nearly all their stuff, so they may be on their way home by now. But, there are many stops and it will take a couple of months probably before they leave, more or less. No one can tell.

Has it rained there yet? This is about the last chance, isn't it. It sure has rained a lot here lately

Write Soon,

Love to all,

Leo

Germany
18 July 1945- **Day 1204** (VE+71days)
Leo H Brown, 90th Infantry, 2nd Battalion, 359th, Company F

Dear Zell and All,

I am faring just fine. Yep, I have shoes now, 2 pair of them, combat boots. The only trouble is that I wear an odd size. They have trouble getting them; it just takes time in the Army, like everything else. There are a lot of people to take care of. I also got an E.T.O. jacket (European Theater of Operations) and two new suits of OD (Olive Drab) clothes now.

I am glad old Fletcher got out of the Army. I actually wouldn't know what to do if I was. It just wouldn't be real. It would be too good to be true. We are supposed to leave here and go home in December. So, I guess I didn't need to worry about getting out anyway.

You must have looked at the wrong town on the map. Delbert and I were 200 miles apart. I sure was glad I got to see him. I was in his town last Saturday but they told me he had left.

I was on a six day tour. We visited all over southern Germany, parts of Austria, and Italy. We saw some interesting sights in the Alps. We also went to Hitler's summer home. It wasn't anything to look at because it had the hell bombed out of it.

I got a package from Cordia yesterday and a letter. Iva was out there then. I wonder if Cordia ever got the divorce. I hope so.

Have you ever had any rain yet? Gosh, I hope so. Mamma said Dennis Lott got enough to plant on but they didn't. Glen and Peet had gone to the wheat harvest. I don't expect the harvest will amount to much. Will it?

The boys put on a parade for General Patton while I was on the trip. He told them it was the best he ever experienced. He thinks a lot of the 90th Division, enough that he will probably take it to the Pacific if he goes.

You said you would like to have heard more about the guy's experiences that had been held prisoner. There was a guy that showed us through one near Munich. He spent $4^{1/2}$ years in German prison under the German SS in that camp. They killed and burned 250,000 people. He didn't mind telling us about it.

Delbert could tell you about it. He saw it (I think) before it was cleaned up. There were over 500 dead bodies that they didn't have time to burn. I saw one near Germany that hadn't been cleaned up but it was not as big scale [sic], though it was much dirtier. The P.W.S. stayed in a cave and worked as long as they could, then the SS killed them and burned them.

Over the door where they burned them was a neat sign, "Happy Ending for Prisoner of War."

Nothing looked happy about it to me. There were 1500 SS defending this one near Munich and over 1000 of them were killed when the Americans took the place. There was an electric fence behind them and they couldn't do anything but fight. They didn't have anything bigger than a machine gun and couldn't do much to our tanks, artillery, etc.

I saw a good movie tonight. I don't know the name of the film; I got there too late to find out.

Write soon,

Love to all,

Leo

Salzbach, Germany
10 August 1945- **Day 1227**
Leo H Brown, 90th Infantry, 2nd Battalion, 359th, Company F

Dear Zell and All,

How does this find everyone? I bet at this exact time you are doing the same thing I am, sitting by the old radio and praying for the war to be over. We get the news every hour. It was really great news for us as well as everyone except the Japs when the first atomic bomb fell on Japan and then the Russians declared war on Japan. It really made us feel happy.

I guess you have seen Delbert by the time you get this.

How is your crop, late I know? Have you had enough rain since it started? I guess you wish you hadn't sold your steers now. But if a person knew what was in the future there wouldn't be any fun living. Would there?

If it hadn't rained when it did I guess Papa would have sold his. Someone wrote me and said Mr. Chapman sold all his cows.

I had a letter from Jimmie Williamson last night. He is still in England.

Write soon,

Love to all,

Leo

Salzbach, Germany
15 August 1945- **Day 1232** (VE day+99 days)
Leo H Brown, 90th Infantry, 2nd Battalion, 359th, Company F

Dear Zell and All,

I just picked up the news less than an hour ago and saw that President Truman said the war is officially over. Thank God.

I am so glad I don't know what to do. It is 2 A.M. now and I am on guard.

I know old Delbert is a glad boy and why shouldn't he be? How this will affect us both, I have not idea but at least we made it through and that is a lot to be thankful for.

Those Japs were so slow I had begun to wonder if they weren't stalling for time.

I bet this is a big day in the States and all over the world.

Love to all,

Leo

France
19 August 1945- **Day 1236** (VE day 103)
Leo H Brown, 90th Infantry, 2nd Battalion, 359th, Company F

Dear Zell and all,

This leaves me just swell. How about yourself? How did Darlene get along with her pony? I suppose Delbert is home now. I hope he never has to go back to the Army.

There is still no news about when we will go home. There are a lot of rumors but no facts. The Colonel said he was sure we would go as soon now or maybe sooner than before the Japs give up.

You wrote about the two towns that Delbert used to be in and the one I am in, I am in Salzbach and not Salzburg. They are a long way apart.

I don't mind telling you I was plenty hot after I read the letter Joe White wrote to the paper. I wrote a letter, with the help of my buddies. We wrote a pretty hot letter to the paper. I hope they print it. We had the personnel clerk look it up in the Army records. The records show that officially the 11th Panzer Division surrendered to the Company G 359th Infantry to Captain Turner who was the G Company commander at that time.

The 11th Panzer was not surrounded. If you remember, the 90th was the first one to reach the Czech soil, so they had to be ahead of the rest of them if we were in contact with them in front. And, no one was up even with us on the flanks. How in the hell could they be surrounded? We were 40 to 50 miles from the Russians. You show me anyone that stayed 24 hours ahead of the Infantry and I will show you a damned liar!

The job of the recon troops is to contact the enemy when they retreat so fast the Infantry can't keep contact. About once or twice that they went out in front of us. You might ask how I know they weren't out in front of us. You might ask how I know they weren't out in front? That is simple! There were always mines and roadblocks that were impossible to get around. There are a lot of ways to tell. We never knew where the roadblocks were and a lot of things we would have known if the Recon had been out there.

Where I usually saw set up was in the same town near the Division C.P. (Command Post) 10 or 15 miles in the rear of the Infantry. When the 11th Panzer came in, they had to come through Company G's outpost line known to us as O.P.L.

We "F&E Company" were holding the main line of resistance known as the "M.L.R." We just put men on each side of the road for about 2 miles. There were men 75 yards apart. When they come in we just started slacking and disarming them. All weapons were still loaded. Tanks still had Ammo in the big guns.

Joe White said they crossed the Rhine River 25 March. I was sure we crossed there on 22 March (Delbert's birthday). But the records show we crossed on 23 March, so the Recon was a little late. I would say the 359th crossed in the afternoon before we crossed early next morning.

Just saw an AM show. I went to church this morning. The church house was packed. I was glad to see that. The Colonel was there

Write soon,

Love Leo

Salzbach, Germany
1 September 1945- **Day 1249** (VE day+116 days)
Leo H Brown, 90th Infantry, 2nd Battalion, 359th, Company F

Dear Zell and All,

How does this find everyone? I am just fine. Things are going about the same over here.

We keep reading in the paper about how many G.I.s that are going home but we still have guys with 95 points that haven't gone yet. I think they will soon now.

They told us yesterday that the 90th Infantry Division would leave for the

States a month earlier that they would have if the war hadn't ended, so that ought to mean some time in November. I don't expect to go home with this Division. I heard I was going to the 11th Armored Division. A guy from that outfit that joined us said they broke it up so I don't know now.

Do you think there will be any crops made this year to amount to anything?

I guess Delbert will be gone by the time you get this letter. I guess he had a big time while he was there. You said he went to Louisiana and I guess I know he had anything to go to Louisiana for. I knew he had a girl in Kentucky but Louisiana was new to me.

No news, so write soon.

Love to all,

Leo

Two Letters From Zell Chapman

Zell wrote often to her Leo and his brother Delbert. Among her items were two letters she wrote to Leo while he was in Germany. The letters were returned. The reader can read an example of the contents of the letters that she shared with him.

Causey, New Mexico
September 5, 1945 **Day-1253**
Addressed to T/Sgt. Leo Brown 38104753, Company F. 359th, APO 90, c/o Postmaster, New York, NY

Dear Leo,

Now since the war is over how are you making it? The same I guess-no difference here but to the ones in combat I imagine its all the difference in the world.

Doubt if I think up much news. I'm poofed out-a bunch of us-Delbert, Verla, Allen, Glen, Eleanor Hay, Peet, Jean& we 4 went roller skating last night and dog my cats if I didn't knock the bottom of the rink out a couple of times. Boy, I'm sore in spots! Ha-They (the kids) all came for supper last night and all of Causey went to Morton to the rink. Think that was about Delbert's first date besides Louisiana.

Can you imagine a guy with a 30-day furlough and wait till the last week to find himself a date? Ha Ha.

Well its dry as can be here. Crops are dying. Guess we weren't supposed to make anything this year.

Am sending off 4 rolls of films of mine & Delbert's this morning-maybe will get them back in a month. I hope-will send some there.

Folks are going to Oklahoma next Sunday when Delbert leaves-The kids and I may go. I haven't decided whether to or not yet. Papa, Mamma, and Jean are the only ones going.

Don is enjoying school except he sure has had a bad arm from his smallpox vaccination.

Guess, I'll quit for now and maybe I'll know some news by the weekend. Write soon.

Love from your Sis.

Zell & Family

Causey, New Mexico
September 20, 1945-**Day 1268**
T/Sgt Leo Brown, 38104755, Co. F, 359th Inf., APO 412, c/o P.M., New York, N.Y.

Dear Leo,

Well at last, I'm getting around to letter writing again-Got two letters from you here while I was in Oklahoma. We got back last Sunday & Cordia had been here since the Wednesday before. Said she came in to surprise everybody and she was the one to be surprised, as Peet & Glen were the only ones in the country. Ha. Cordia came over and spent Tues, night [sic]& Wed. with me & I've been ironing 3 days straight to get over, as at last I've found a minute even if I am in the bed retired for then night, I mean.

Cordia lost her job & she and Leveda came in on a 2 weeks visit or 12 days I think& then they're going back, to what they haven't decided yet-I don't think. Leveda is clerking in a grocery store and I think Cordia said she was going to try a Beauty Course & get her a license to operate a shop in about 6 months. She got her divorce about a week or so ago. She, Delbert and Iva just missed one another all the way around.

You asked about Delbert and Louisiana. His Kentucky girl lives in Baton Rouge and has for quite awhile-since Delbert shipped out I think. She works there. He met her dad, a 28-year Army man-been in 28 years I mean. Quite OK for Delbert I think OK for Pa-in-law-I guess. Ha, Ha, Ha.

Got your letter telling about the letter from the paper I sent you that Joe White wrote. Folks didn't want me to send it to you, but, I said if someone were trying to make me out a liar, I'd want to know it, so I sent it on to you.

Saw Oscar Scott Monday. He'd got his discharge and said Delbert went on to the Camp Cook, California. Heard today where MacArthur is only asking for 2 more Divisions instead of 17 as was previously planned-Surely Delbert wont hit one of those 2. Also said in the news today any man who had been in Army 2 years would be released before spring, so I guess he'' hit that anyway-no doubt you'll be discharged if you get back here. Harold Todd is discharged now too- Notice where you told some of the folks no to send Christmas packages. I was fixing to send one but will wait now.

You asked if there would be any crops made. No, I don't think. Everything I've seen is fried-about half in fact and some so dry it rattles so you know a foot of rain wouldn't help that. We looked our crop over today and its ruined and the land is so loose and dry and clean that it is beginning to blow. Some have kept their land ready for wheat and it is drifting like March. Sure wish we could get a good rain to pack the ground if for nothing else. We have stuff to hold our land. I think made a mess of black-eyed peas for dinner today if we don't make any crop. Ha, Ha.

Enjoyed my trip to Oklahoma O.K. but I'm more convinced than ever that its not the place for me. They had no crops because they washed out till too late to plant-some have 10 or 12 acres of corn left. Weeds are higher than your head everywhere and Johnson grass I never seen any thing like it. As Allen always said, they are crowded there already and I don't plan to crowd them any worse. Ha. Ha.

Guess Delbert was a blue puncher when he had to go on to camp-several that went back to Fort Bliss when he did were back here in a week. He told Allen if he seen that he'd get out in time for a crop next year he wished he'd be on the lookout for him a farm-I didn't know whether he'd farm again or not.

It's blowing-thundering& lightening now but that don't mean rain I've already decided.

Guess I better close and get some shuteye. Be glad when we get back on standard time. 10 o'clock won't sound as late as 11 to got to bed tho they're the same on (war) time as 11 as 10 will be there.

Excuse the pencil, but I decided I'd not try an inkbottle up here in bed. Ha.

Write soon-

Love Ever

Your Sis & All

Zell

P.S. I didn't tell you who all we seen on our trip. Saw Uncle Jim & folks-Trudies, Vertas, Millie, Minnie, Aunt Kate, guess you may remember her-Uncle Bud's folks and Mama's uncle Marshall Anderson. Saw a lot of Uncle Bud's married kids and Aunt Millie's too. To many to mention and I hardly remembered any of them (dumb-no)

This letter was on Red Cross stationery and was written from Camp Lucky Strike, Leo's final assembly location before his journey home.

France
26 September 1945- **Day 1274** (VE day+141days)
Leo H Brown, 90th Infantry, 2nd Battalion, 359th, Company F

Dear Zell and all,

How does this find everyone? I am O.K. I am at Camp Lucky Strike about 40 miles from Le Havre, France. I don't know a thing about when we will leave. There are some guys here that have been in this place for two months over the High Point Men that left the Company on 8 August that are still here.

We might leave here in a couple of weeks or a couple of months. No one knows, seems like.

Say Zell, I sent a box full of junk to your address. I wish you would save it for me. It ranges from a clipping out of the paper to a piece of rock I broke off of Hitler's fireplace. (A good chance that a piece of Hitler's fireplace is still on the ground somewhere in Roosevelt County, New Mexico)

Do you think there will any crops made in that country to amount to anything this year?

There isn't any use to write to me for no mail comes into this place. I haven't had any mail since I left the Company. It is a long way from Czechoslovakia in one of those 40 of 8 boxcars. I had rather be on the move than sitting still. Time passes quicker that way. It rains every day and night here. It is really muddy.

Love to All,

Leo

OCCUPATION
PREPARING FOR JAPAN

There were a lot of people who had fled their homes, but a lot had also stayed. Next door to us, there was an old blacksmith. He had been forced to work for the Germans for several years in which he had to build wheels for their horse-drawn equipment. He was doing all of this by hand. He could speak a little English and I could speak a little German.

It took me a while, but he finally realized that he didn't have to work for them anymore. So, he sat on his anvil and we talked what little we could. He had a long-stemmed pipe that reached down as far as his belt. He got a handful of wood shavings, ground them up in his hands, and was about to put them in his pipe when I went to the kitchen truck and got him a new can of pipe tobacco and he loaded up and lit his pipe. I don't think I ever saw anyone as happy as he was.

When we left the next day, someone said that the town's people built an arch over the road out of town and it was built out of flowers. Everyone had on their Sunday's best clothes.

We moved back to Germany and occupation duty. We moved to

a village that was located twenty-two miles east of Nuremberg. It had some old German barracks that were pretty crude, but they were dry. We managed to rig up a shower. We found some good wheat-straw for bedding. It was too good to be true! No one was shooting at us. We were assigned occupation duty.

The War Department set up the point system. It worked as follows: one point per month for the time put in the States, two points per month for overseas duty. Ten points for a Purple Heart, ten points for a Campaign Star, ten points for a Combat Infantry Badge, and ten points for any action of honor such as the Silver Star, Bronze Start, etc. If the soldier was married, each child was worth ten points even if they were born after the soldier had left for war.

We had heard that high-point were men getting discharged as soon as Germany surrendered. I suppose that I didn't have a worry that I would be one of these high-point soldiers. The required points to be discharged early was eighty-five. I couldn't believe that I only had seventy-six points; nine short of being discharged! As long as the war with Japan went on, there wasn't any way to get any more points.

We did send out several high-point men. One was old Stanley Skay. He wasn't in the hospital long enough to miss a Campaign Star. He had a Purple Heart and two Clusters, so he had plenty.

We did a lot of guard duty, patrols, and parades. We were the middle company of our regiment, and I was the enlisted man with the most combat experience, so I carried the flag.

The reason that we did so much patrolling and searching houses and had so many rules to follow was that it was forbidden for the Germans to have more than five people assembled. The G2 people got word that the S.S. troops were going to wage a guerrilla war in the hills of Southern Germany. We never found any sign of anything except total surrender.

By the time I left Europe, we had released all of the regular armies and were holding S.S. troops. War prisoners were working on cutting wood and repairing damaged houses, as well as many other tasks. Some of the people, however, just quit and thought that we were going to take care of them. Some of them wouldn't harvest their crops or work in the fields. They thought the "American Sheriff' would feed them. One local woman asked me, "When is the sheriff going to rebuild my house?"

I answered, "It might be a while."

The people seemed to be a beaten and conquered nation, and they accepted it.

One night, before the war was over, we were staying in a town. A lady came up to me and asked, "Will there be any shelling of our town tonight?" She wanted to sleep in her house. She said, "We have had to sleep in the basement for two months."

I told her, "The Germans might shell it, but the Americans sure won't."

She replied, "Oh, they are finished. They can't do anything."

Not long after this, a man came to me and wanted to know, "Who is the sheriff?"

I told him, "I am the Sheriff!"

He replied, "A shell hit my barn and killed two of my hogs. May I have permission to butcher them?"

I said, "Go ahead. It will be okay." He couldn't believe his good luck. The first person he asked was the "sheriff," and his request had been approved immediately! He and some others helped him pile dry wheat straw on top of the hogs, and they set the straw on fire. After the hay had burned out, they turned the hogs over and repeated the process. Then they hung them up, threw water on them, and began scraping the skin. They were white, without a scratch or a blister! I thought they would be half cooked.

After the war and V.E. (Victory in Europe) Day was over, we were still settled down in a little town in the old German barracks. A lot of guys were letting people do their laundry. That had always been a problem, except in Fort Dix, New Jersey. Anyway, I just didn't trust the locals. I don't suppose they liked us too much. I thought that if I let them have my clothes to wash and never saw them again, I would have a problem.

One night, as I was coming from chow and it was getting dark, I passed an evergreen tree at the corner of the building. As I passed it, there was this wee little voice that said, "Mister, would you like for my mother to wash your clothes?"

Well, I did want them cleaned, but I didn't trust him or others like him. His mother was behind him back in the branches of the tree. She had him out in front because he could speak English better than she could. I asked him, "What do you want to wash my clothes?"

They had a little talk and he said, "We want a penny box of matches."

They didn't want money. They couldn't buy anything with money, nor could they cook anything without fire. Cigarettes became the primary medium of exchange. I went and got my clothes and gave them to her. She and her son didn't expect anything until they returned the clothes. I gave her a few matches and two cigarettes, and I gave her little boy some hard candy. I promised them some more when they returned my washed clothes.

The next night, my clothes were back and washed with a brush in cold water. They sure wouldn't last long that way, but they were spotless. From then until I left, the little boy picked up and delivered the clothes. I was much more generous to him on delivery. They were always prompt and did an excellent job. The little boy liked the hard candy. I didn't give him much when he picked up the clothes, but I was much more generous when they were returned. I don't know how many people they were washing for, but they never got the clothes mixed up. I would hate to think our women would have to wash enemy soldiers' clothes for a penny box of matches or any price.

During this time, we had the opportunity to take a field trip to Southern Germany, Austria, and Northern Italy. We started from

Regensburg to Nuremberg; we saw the stadium where the Germans had held their big military parades. Hitler would scream and yell about how they would rule for a thousand years! It turned out this was a fairy tale. The Allied forces virtually destroyed Nuremberg and Munich during the aerial bombing.

We crossed the Danube, traveled to Donauworth, and went on to Augsburg (founded in 15 A.D.), where the first Protestant church was located. Munich was where Hitler had a beer hall, a monument to Nazi martyrs, and a memorial to Bavarian kings. From Munich, we went to Oberammergau and Linderhof, where we saw Ludwig's palace. We spent the night in Innsbruck, which was a beautiful ski resort. We went for a tour on the cable car. It was rusty and pretty out of shape for my comfort. Getting off the thing was good!

From there, we went to the Dachau concentration camp. It had already been evacuated and cleaned up. Our guide was a young Jewish man who himself had been on the list to be executed a few days after the Americans liberated the prisoners. He seemed to feel very indebted to the American Army. He said, "I have been born twice; the first time by my mother, the second time by American soldiers."

We went to Berchtesgaden, Hitler's home, and saw the notorious "Eagle's Nest." What a strange experience! We continued our tour with stops in several other locations, including Hitler's birthplace, Braunau. We crossed the Danube again and returned to the place where we had started.

By now, we were operating like a peacetime army. Everything

was according to the T.O. (table of organization). We had to sign for everything, and the supply sergeant didn't let me forget to sign for the G.I. watch. I passed it on to Sergeant Carson, who took over when I left. He took it and signed for it so the supply sergeant couldn't get it. He said, "One way or another, I will see to it that the supply sergeant doesn't get it!"

We had to equip, according to the T.O., three motor squads and two machine-gun squads. Most of them were not trained to do that job, but the war was over and it all worked out. Our occupational duty turned into training to go to Japan. We were training ten hours per day for combat with Japan.

The fellow who deserted us back on the Saar River was to be tried, and I was to be a witness. That is because I was the one who sent him back to the aid station. There was a strict fraternization law. This soldier was under house arrest, and Captain Lesser caught him with a German girl and arrested him. The defense attorney, a first lieutenant, wanted me to say that it was night and that he could have been lost. He and Pierce (the one in trouble back at Camp Barkeley) were arrested and sent back to stand trial. They broke arrest and took off again. The defense attorney at the court marshal told the soldier that he would draw some time, so he had just as well take all of the charges at once.

The prosecutor, the Provo Marshal, read the charges. The soldier pleaded guilty of breaking house arrest but not to fraternizing. The defense attorney asked Captain Lesser, "Did you catch the man fraternizing?"

Captain Lesser answered, "I did."

The attorney continued, "Then you placed the girl under arrest?"

He replied, "Well, no. I had no right to place her under arrest."

The attorney then said, "You did look at her papers to determine if she was German?"

Captain Lesser said, "No, I did not."

The attorney said, "How do you know she was German? Could you swear that she wasn't Russian? In fact, could you swear she wasn't an American girl?"

Captain Lesser said, "I can't."

They dropped charges, and the soldier drew six months time. After serving six weeks, I got a letter from the Provo Marshal, and it asked if I would take the young man back and be responsible for him. If I did, they would parole him to me. I replied to the Provo Marshal's letter and told him that I thought the soldier got an excellent deal and that I just didn't want him back. Besides, my men wouldn't accept him after what he did. I never heard another word from the Provo Marshal.

Pierce didn't get tried at this trial. Both men had deserted and had gone to France. They were arrested in France, then sent back to the F Company to be tried for desertion. Both broke arrest and deserted again, but they were arrested a second time. Pierce was later tried and got sixty years in prison, which was to be carried out in France. All of the court marshals were reviewed after the end of the war, and most of the sentences were reduced. I hope Pierce had his sentence reduced. He wasn't a bad sort; he just got off on the wrong foot and didn't do anything to correct it.

Our occupational duty turned into training to go to Japan. There was a mad scramble with officers to get a promotion before combat ended with Japan. Soon, it would be a peacetime army, and then promotions would be hard to come by. Our fearless second lieutenant was promoted to first lieutenant because he was in combat for more than sixty days. He got to be company executive officer and took the battalion commander post. A lot of the old guys were shipping out.I couldn't see how we were going to put together an army to send to fight the Japanese and win. It bothered me. However, we were saved by Japan's surrender. I was offered the position of first sergeant. I turned it down. I think it was a job and no promotion.

We continued doing some training but not going about it seriously. I could have gotten into trouble one day during a training exercise. A jinglebob lieutenant colonel had told me to meet him at a particular place for a training exercise. The place was about one-fourth of a mile from a building with no doors or windows on our side. There was a flat white gravel road to the back of the building. To our right was an outcrop of huge boulders. The lieutenant got us all spaced out and said, "Fix bayonets! And charge the back of the building and scream at the top of your lungs!" The rest of the men took off, and I took about half a dozen steps, took off my helmet, and leaned my rifle against a boulder and sat down.

About that time, a major, the new battalion commander, drove up and said, "Sergeant, is this your platoon?"

I said, "Yes."

He said, “But you are not involved with it!"

I wasn't too happy and said, "If this is the way we do things, I won't be involved with it! In real combat, not one of those men would have reached the building alive."

He said, "You don't think so?"

I answered, "I know that two machine guns will cut every man there down."

I replied, "How do you think you would have pulled it off?"

I said, "If I had to attack the building, I would put my men out in the boulders before daylight and would use most of my machine-gun ammo and drop mortar fire through the roof. Then, I would charge the building at daybreak instead of broad daylight."

The major said, "Sergeant, your idea has merit."

He drove off, and we never had any more training exercises. Then, the big bomb-of-all-bombs (the atomic bomb) was dropped on Japan and, after a week, all training exercises were dismissed.

My platoon was on guard duty, and I was listening to Armed Forces Radio. It was 2:00 a.m. Everyone who wasn't on guard duty was asleep when the news came that Japan had accepted the terms of surrender. I turned all the lights on and was yelling. Everyone thought I was crazy until they found out what the deal was and then they joined me.

One Saturday morning we had a parade ground inspection and our brilliant lieutenant came to be sure he knew how to pull off an inspection. About that time a very drunk soldier (I think he was Indian) came in and tripped over the doorsill and nearly fell in the lieutenant's lap. He jumped up and said, "Sergeant Brown this man is drunk!"

I said, "I will agree with you on that one."

He then said, "He can't stand inspection like this!"

I replied, "Right again. We'll have to leave him here. There isn't anything else to do."

This soldier was one of the replacements we got out of the Air Force, and we were just then finding out why. He started ridiculing the lieutenant. Most of these Air Force guys had been in a long time, and the soldier was reminding the lieutenant of that fact. We left "Chief" (that is what all the men called him), and about that time we lined up for inspection. Here came Chief with all of his mess gear and it was rattling. I told one of his buddies whom he had been in the Air Force with, "Take care of him!"

The battalion commander (the major) asked me, "Why is that soldier breaking rank?"

I said, "One was on sick call, and I had to replace him."

He said, "Oh, I see (he didn't)."

It worked out.

I wrote about Red Cadwell, the soldier from Borger, Texas, getting killed. On his rifle stock, he carved the following: "Red Cadwell, Borger, Texas." Someone had taken his rifle. The major said, "Sergeant, as soon as this inspection is over, I want to see this man who carved his name on this rifle."

I said, "Sir, that wouldn't be possible because he was killed over a month ago."

I would like to have said, "Because you got him killed!"

He probably never gave a second thought to carving his name on his rifle.

When we got back to the barracks, I asked one of the fellows how he managed to get Chief to go back with him. He said, "It was simple. I told him I had a little stash in the attic and he came."

While we were waiting for transportation, a bunch of us were offered to go on a tour of Austria and Italy. We took our sleeping bags. We ate with various units along the way. It wasn't a luxury tour, but I am glad we got to make it.

After V.J. (Victory in Japan) Day, I had plenty of points, and so did a bunch of other guys. The 16th Armored Division in Czechoslovakia had started taking high-point men to go home, so I started the long road home. There were several out of our company, but the only person that

I remember was John Davis. We were officially part of the 16th Armored Division, but we just checked in, stayed there for a few days, and then we checked out and boarded a train.

When we got to the 16th Armored Division in Czechoslovakia, they told us we could roam around anywhere we wanted to but were told not to cross the road north of us because it was Russian territory. They said that the road was neutral, but if we were to cross it, we would be on our own. Davis and I decided to look around a little. We got out on this road and were walking up this fairly steep hill when an old truck passed us, going slow. We just stepped up on the back of the truck and, when it started down the hill, it was going much too fast to get off of, and we could see this Russian checkpoint with a log across the road. It looked like trouble.

The truck stopped, and these two tall Russians came up to us, gave us a big salute, took us by the arm, and we crossed into their territory to a beer joint. They bought us a beer. I knew John Davis could speak some of their Slovak languages, so we drank the beer and he started in on what trouble we were in with our commander if we weren't back at a certain time. So, we gave them all the cigarettes we had, and they took us back over to our side of the road and gave us another salute and a handshake. We then took off. We walked fast until we got out of sight; neither of us saying anything, but running about a mile before we slowed down. Things could have been a little different and we could have been in big trouble.

On our way back to Germany and France, we rode in the forty

and eight boxcars, but we only had twenty men to each car. We would stop along the way and eat at transient mess halls. One night, we stopped a little before dark and ate, and then loaded back up and went on. At breakfast time, someone was beating on the door and yelling, "Chow!" I didn't seem all too hungry, but we never refused chow; we had things that didn't normally go for breakfast. Someone looked at his watch, and it was 9:30. We had eaten someone else's supper! The cooks were frantic, but we couldn't give it back. I felt bad about that, but there was nothing we could do.

We landed at a tent city near Le Havre, France (on the coast). It was called "Camp Lucky Strike." We were there waiting on a boat and, who did we see, but old Stanley Skay and a bunch of those high-point soldiers that left us in May, and now it was October.

GOING HOME

We were ready to load up, as they were just getting there. We were loading up on a new Victory Ship the next day. There was a captain just in front of me. He had an officer's bag all stuffed. I had a few clothes and personal items in the bottom of my duffel bag. The captain asked, "Would you give me a lift?" I did, and he handed me the white sailor's bag. It was a rather long peer with only a 2X4 rail on each side. I was kind of kicking the bag along as we moved up and he said, "Get a better hold on that. This is the company payroll, and it has 90,000 dollars in it!" That was a lot of money in those days.

The chow was good. It was a good ship, but it was small and rolled and tumbled in the North Atlantic; being a new ship there didn't enable a system for serving meals. You could put in about as much as you could stand waiting in lines. I put in one night on K.P., and I figured a system to eat. I had washed the trays that night. From then on, I would roll up my sleeves, get a stack of trays, put them at the end of the line, keep the bottom one, and go through the line. It worked like a charm.

I got sick coming home, and a cook gave me two lemons and said, "Eat one and save the other one. “If you feel sick, suck on it." It worked.

The return trip took seven days. When I heard the anchor chain running, I jerked on my britches and shoes and ran up on deck. That was a beautiful sight: Boston Harbor! There were the lights of the good old U.S.A. on all three sides.

After debarking from the ship, we stayed in a little boot camp made out of sheet rock and tarpaper. The next day, Davis and I were walking around camp and decided we would get a haircut, a hot oil shampoo, and the works. We did, and the price was 7.50 dollars (that was a lot of money in those days). I never thought anything about it because I didn't ask in advance, but Davis came unglued and threw his wallet at the barber's feet and said, "Take it all, you money hungry S.O.B.!"

I finally picked up the wallet and paid the man and got Davis by the arm and led him out. I had never seen him do anything like this, and I had known him for three years. The next day, we split up, and I left. Names were called out for train cars and the highest ranking N.C.O. (enlisted man) was in charge and responsible for seeing that all the soldiers got to where they were going. I just knew it would be me, and it was a long way to Fort Bliss, Texas, but I got lucky because there was a master sergeant in our group.

His name was "Outlaw." He was from Arizona. We went down the eastern seaboard, dropping off a train car every so often. When we got to Dallas, Texas, our two military cars were hooked to a civilian train. We had quite a wait. I don't know where all the whiskey came from, but Outlaw and I were the only ones who didn't have a fifth of whiskey.

We took on a couple of new lieutenants as transportation officers, and they came in and said, "We'll be back in one hour. In that time, throw all of the whiskey out." An hour later, there wasn't any whiskey to throw out. Outlaw and I were the only ones sitting up. Sometime about morning, we arrived in El Paso, Texas, and Fort Bliss.

We got a lot of checking and exams for nearly a week. Finally, I was given my discharge early on the morning of October 26, 1945. I was given $14.68; the price of a ticket to Portales, New Mexico.

I took the bus to El Paso and bought my ticket to Portales with my $14.68. On our way through the Hondo Valley, one of the rear breaks wore out and fell off of the bus. We were to change buses in Roswell, New Mexico, but there wasn't one to change to, so we waited (I was used to waiting). Finally, late in the afternoon, we left.

Somewhere around Elida, about thirty miles from Portales, I remembered that the next day was Sunday and we always fixed the windmills on Sunday.

I arrived in Portales at around 8:00 p.m. There was a barber's shop joining the bus station. I had a shave early that morning and a shoeshine, but I got another shave and shoeshine. I saw a guy on the street that I knew, and he said, "I know where your brothers are." Well, that was a relief because I was still thirty miles from our farm.

There was a little dance hall out on West 18th Street. When I got there, I walked in to locate my brothers. A guy who was selling tickets stopped me and said, "It will cost you $1.50 to dance."

I told him, "I had been gone for nearly two years and am not interested in dancing. I want to find my brothers and go home now since I am this close."

He said," Buddy, if you get past me, it will cost you $1.50."

Of course, I should have paid him, but I just didn't feel like it. I said, "You just watch me!" I walked by anyway and nothing happened.

We loaded up and went home. The folks got up, and I kissed my mom on the cheek and shook my dad's hand. He said, "Son, it is good to have you home. The windmill is broken, and we will have to fix it tomorrow. On the following day, we fixed the windmill.

It was a long road home, but it could have been a lot worse.

Leo Brown

P.S.

Shortly after I got home, I wrote Odell Sullivan and learned that he had been killed in a car wreck a couple of weeks earlier. The only person mentioned in this paper that I still have contact with is my old and dear friend, O.D. Desenberg.

P.S.S.

I let a few days go by, but I had to see Mrs. Mitchell, as I had told her I would. Homer's body had not been recovered, and so he was listed as missing in action (M.I.A.) for one year. I thought of (but not seriously) letting her hang on to the hope that he might one day show up. However, that would have been cruel when there was no chance of it happening. So I said, "Mrs. Mitchell, I will tell you the truth to all that I know."

She said, "That is what I want." She continued, "Is Homer dead? Do you know for sure?"

I said, "Homer is dead, and I saw him. He isn't coming home."

She didn't shed a tear, but said, "This I can get over because I have been sure of it the whole time. As long as there was hope, we hung on to it."

I later got word that the Bryants wanted to talk to me because I had seen and spoken to their son, Cleo Bryant, shortly before he was killed. They seemed to think that we were buddies and knew each other well but that was not the case. A division is big, and we weren't even in the same division.

They invited my folks and me up for a Sunday dinner. We ate and then went into the living room. Mr. Bryant said, "Now tell us about Cleo."

I felt about two inches tall because I had met him by accident and found out that we were both from Roosevelt County. I had probably spoken to him for fifteen minutes. The next time I had heard about him was when I got a letter from his sister wanting to know how he was killed and anything else I knew about his circumstances.

It just seemed to be very important to his father to know how he was killed. I think that he more or less looked at the war as if it were a western shoot-out. He asked, "There wasn't any chance that the German was the best man or fastest on the draw?"

I told him that there was about a ten-to-one chance that he was killed by artillery. But if he wasn't, it wasn't a personal thing on a one-to-one basis. That seemed to ease him some.

History

The following is not from my memory. It is the history of the 90th Division written shortly after World War II by First Lieutenant Joe Abrams, Division Historian. In the European Theater of War, 78,000 American men were missing or unaccounted for. I am sure that many of them were later listed as K.I.A. (Killed in Action) after one year, which was the requirement.

The 90th Infantry Division Total Casualties

Treated by all Medical Units 25,988

(M.I.A.) 2,260

(K.I.A.) 3,340

Leo's letters home to his sister Zell in Causey, New Mexico

Delbert W. Brown and Leo H. Brown: Reunion 1945. (The brothers reunited in Europe after the Battle of the Bulge. Delbert told his sister that Leo looked like a "gutted snowbird."

Sgt Leo H. Brown. Notice the dog tags that were with him for his entire stay.

ABOUT THE AUTHOR

Leo H. Brown died peacefully in 2006 in the V.A.M.C. Veteran's Hospital in Amarillo, Texas with his wife, daughter, grandchildren, and son-in-law by his side. His brother Delbert and his grandson Aaron came to pay their respects. His watch had ended and in the words of the Apostle Paul of the New Testament, "I have fought the good fight, I have finished the race, I have kept the faith. Now there is in store for me the crown of righteousness, which the Lord, the righteous Judge, will award to me on that day—and not only to me, but also to all who have longed for his appearing." –2 Timothy 4:7-8

R.D. Jentzen is a fulltime medical provider, enjoys travel, reading, and cycling and listened to Leo's stories for nearly three decades.

Made in the USA
Monee, IL
04 November 2026

67385511R00154